Magnetospheric Physics

Achievements and Prospects

Magnetospheric Physics

Achievements and Prospects

Edited by
Bengt Hultqvist
Swedish Institute of Space Physics
Kiruna, Sweden

and
C.-G. Fälthammar
Royal Institute of Technology
Stockholm, Sweden

Springer Science+Business Media, LLC

Library of Congress Cataloging-in-Publication Data

Crafoord Symposium on Magnetospheric Physics: Achievements and Prospects (1989 : Stockholm, Sweden)
Magnetospheric physics : achievements and prospects / edited by Bengt Hultqvist and C.-G. Fälthammar.
p. cm.
"Proceedings of the Crafoord Symposium on Magnetospheric Physics: Achievements and Prospects, held September 28-29, 1989, in Stockholm, Sweden"--Verso of t.p.
Includes bibliographical references and index.
ISBN 978-1-4615-7378-4 ISBN 978-1-4615-7376-0 (eBook)
DOI 10.1007/978-1-4615-7376-0
1. Magnetosphere--Congresses. I. Hultqvist, Bengt. II. Fälthammar, Carl-Gunne. III. Title.
QC809.M35C73 1989
551.5'14--dc20 90-21579
CIP

Proceedings of the Crafoord Symposium on Magnetospheric Physics: Achievements and Prospects, held September 28-29, 1989, in Stockholm, Sweden

ISBN 978-1-4615-7378-4

Originally published by Plenum Press, New York in 1990
Softcover reprint of the hardcover 1st edition 1990

PREFACE

This book contains the proceedings of the 1989 Crafoord Symposium organized by the Royal Swedish Academy of Sciences. The scientific field for the Crafoord Prize of 1989 was decided in 1988 by the Academy to be Magnetospheric Physics. On September 27, 1989 the Academy awarded the 1989 Crafoord Prize to Professor J.A. Van Allen, Iowa City, USA "for his pioneer work in space research, in particular for the discovery of the high energy charged particles that are trapped in the Earth's magnetic field and form the radiation belts - often called the Van Allen belts - around the Earth". The subject for the Crafoord Symposium, which was held on September 28-29 at the Royal Swedish Academy of Sciences in Stockholm, was Magnetospheric Physics, Achievements and Prospects.

Some seventy of the world's leading scientists in magnetospheric physics (see list of participants) were invited to the Symposium. The program contained only invited papers.

After the presentation of the Crafoord Prize Laureate, Prof. J.A. Van Allen, and his specially invited lecture: "Active Experiments in Magnetospheric Physics" follows in these proceedings two papers on the achievements of magnetospheric research hitherto. The main part of the proceedings (8 papers) deal with the main theme of the Symposium: How we shall carry on magnetospheric research in the future.

The Symposium was organized by five members of the Academy representing the field of space physics: Lars Block (Stockholm), Rolf Boström (Uppsala), Kerstin Fredga (Stockholm), Carl-Gunne Fälthammar (Stockholm) and Bengt Hultqvist (Kiruna, Chairman).

The Symposium was financed by the Crafoord Foundation through its donations to the Royal Swedish Academy of Sciences and contributions were also given by the Swedish Board for Space Activities and the Swedish Institute of Space Physics.

We owe many thanks to staff members of the Royal Swedish Academy of Sciences, the Royal Institute of Technology and the Swedish Institute of Space Physics, who contributed in important ways to the organization of the Symposium and to the production of these proceedings, in particular (in alphabetical order) Stig Björklund, Eivor Jonsson and Margaretha Wiberg.

Bengt Hultqvist   Carl-Gunne Fälthammar

CONTENTS

PRESENTATION OF PROFESSOR JAMES A. VAN ALLEN AS THE CRAFOORD PRIZE LAUREATE OF 1989 AT THE ROYAL SWEDISH ACADEMY OF SCIENCES, SEPTEMBER 27, 1989 (by Bengt Hultqvist)

Your Majesties, Your Excellencies, Mrs Crafoord, Mr President, Ladies and Gentlemen:

The Royal Swedish Academy of Sciences has decided to award the Crafoord Prize 1989 to Professor James Van Allen for his pioneering exploration of space, in particular the discovery of the energetic particles trapped in the geomagnetic field, which form the radiation belts - the Van Allen belts - around the Earth.

Fig. 1. Professor James A. Van Allen

Magnetospheric Physics, Edited by B. Hultqvist and C.-G. Fälthammar
Plenum Press, New York, 1990

The Van Allen belts can be seen in Fig, 2, deep inside the magnetosphere. They consist of energetic charged particles - that is electrons and ions - with very high energies (up to several hundred MeV) i.e. the highest energies found in the magnetosphere. They move around the Earth in the geomagnetic field (as shown in Fig. 3) with continuous bouncing between mirroring points in the two opposite hemispheres and with a circular motion around the magnetic field lines as the most short-period of these three periodic components of the motion.

The discovery of the Van Allen belts in 1958 was the first great surprise of space research. The Norwegian physicist Störmer had shown theoretically many years before that charged particles can move in trapped orbits around the Earth, but as such orbits can be reached from neither outside nor from inside in a static system, no one believed that there could exist particles in such orbits, at least not in large numbers. Van Allens's Geiger-Muller tube onboard the first US satellite, Explorer I, showed a strongly decreased counting rate at altitudes above 1000 km in a region in the South Atlantic, where the geomagnetic field is weaker than over other parts of the Earth. Van Allen interpreted the low count rate as due to the particle flux being so high that the instrument was overloaded and saturated. This interpretation was confirmed by his later investigations during early 1958.

The Van Allen belts are not only important scientifically but they also affect practical space activities around the Earth. The practical effects have primarily to do with the high radiation intensity in them. The fluxes of energetic particles are so high that manned spacecraft have to stay out of them or pass through them as fast as possible, and even for unmanned spacecraft the orbits are generally chosen so that the radiation belts are avoided as much as possible in order not to have the electronic equipment damaged. Radiation doses of 10^5 rad (10^3 Gray), above which it is difficult to keep electronic equipment alive, are easily obtained in satellite orbits which spend much time within the more intense parts of the Van Allen belts. Only a few thousandths of these values, i.e. a few hundred rad or a few Gr, are deadly doses for human beings.

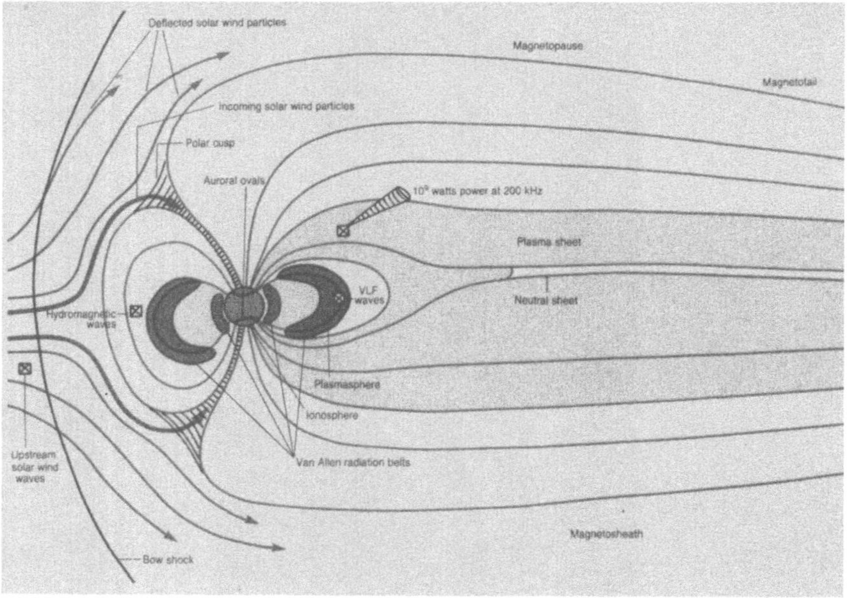

Fig. 2. The magnetosphere of the Earth with the Van Allen belts marked (after Lanzerotti and Krimigis, 1985).

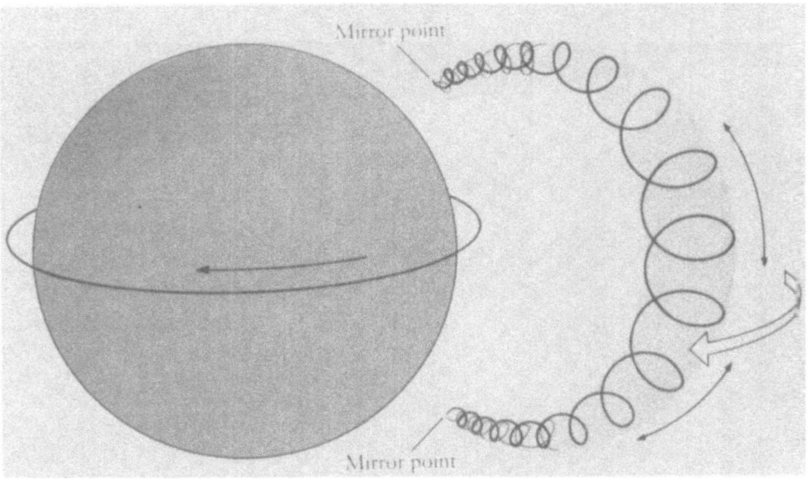

Fig. 3. The three periodic components of charged particle motion in a magnetic dipole field: gyration around the magnetic field lines, bouncing between the mirror points in the two hemispheres and drift motion around the magnetic dipole.

Fig. 4. The first US satellite, Explorer I, with the outer shell and the nose cone removed and with the Geiger Müller tube at the center.

Fig. 5. Professor James A. Van Allen, at the center, at the press conference on the 1st of February 1958 after the successful launch of the first American satellite, Explorer I. To the right of Van Allen is Wernher von Braun, who was responsible for the rocket, and to the left William Pickering who was in charge of telemetry reception, tracking and other ground based activities.

Van Allen's instrument for energetic particles was the only scientific instrument on Explorer I, besides a small micrometeorite detector. Fig. 4 shows Explorer 1 with the outer shell and the nose cone removed and with the GM tube at the center. In Fig. 5 James Van Allen is seen, at the center, at the press conference on the 1st of February, 1958 after the successful launch of Explorer I. We see him together with Wernher Von Braun, at right, who was responsible for the rocket, and William Pickering who was in charge of telemetry reception and tracking and other groundbased activities. They are holding a full scale model of Explorer I. It was not a large satellite. The discovery of the Van Allen belts resulted among other things in that Van Allen also became cover boy on the Time magazine, as shown in Fig. 6.

These successful satellite measurements in early 1958 were the result of a long period of scientific, technical and organizational preparations. In the 1950s Van Allen was the leader of sounding rocket programs, studying cosmic rays, aurora and the geomagnetic field. In the period 1952-1957 rocket launches were conducted with the help of balloons from ships. The rockets were carried to high altitudes with balloons before being fired. In all 57 such rockets were launched successfully. Van Allen also led the development of the Aerobee sounding rocket for scientific use. A large number of them have been launched through the years. He was for many years chairman of the Upper Atmosphere Rocket Research Panel which was the focal point of all space research in the United States before NASA was formed in 1958. His superior expertise made him uniquely qualified as principal investigator on the first US satellites.

James Van Allen's experimental research in space started when he became a professor in physics at the University of Iowa in 1951. Going to Iowa meant that he returned to the state where he was born and educated. He had received his doctor's degree at the State University of Iowa in 1939. The years before 1951 he spent at the Carnegie Institution in Washington DC, at Applied Physics Laboratory of Johns Hopkins University and in the US Navy during the war. At the University of Iowa he built up a space research group which became the leading university group in space physics in the country. Able graduate students gathered around him. In Van Allen's group at the University of Iowa a larger number of prominent space physicists of the first generation have been educated than at any other university. Many of the early students are today the leaders in space physics research in different parts of the country. We see some of these first students, who have later become famous themselves, in the next two figures. In Fig. 7 Carl McIlwain (left) and George Ludwig (right) are seen together with Van Allen and in Fig. 8 Lou Frank, who has continued Van Allen's particle measurement program at the University of Iowa. We are happy to have two of these three of the first Van Allen students with us here today, namely Carl McIlwain and Lou Frank, as well as several other students of Van Allen's: Dan Baker, Ted Fritz, Don Gurnett, Tom Krimigis and Stan Shawhan.

Fig. 6. Professor James Van Allen on the cover of the Time magazine in May 1959.

Fig. 7. Carl McIlwain (left) and George Ludwig (right) together with their teacher, James Van Allen.

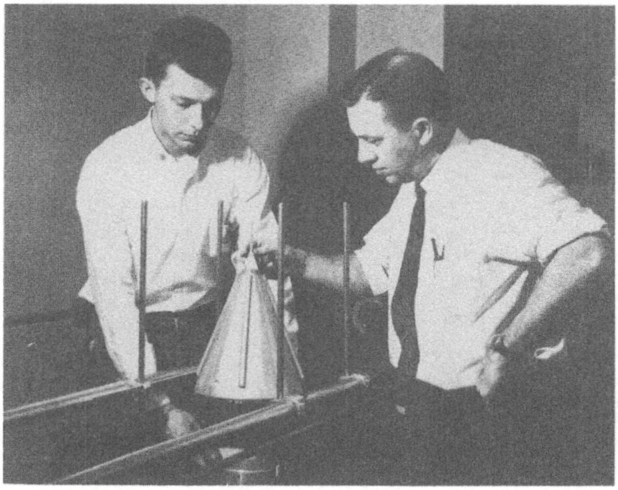

Fig. 8. Louis Frank, who has continued James Van Allen's energetic particle measurement program at the University of Iowa, as a young student together with his teacher.

The work pace was very hard for Van Allen and his small group in 1958. There was one satellite launch per month in the early part of the year. The rocket that should bring Explorer II into orbit failed, but Explorer III was launched successfully already on the 26th of March, i.e. less than two months after Explorer I. Explorer IV was sent up on the 26th of July, i.e. four months later, and the Van Allen group supplied scientific instruments for two additional missions in 1958, namely Pioneer I and Pioneer III which were sent to impact the moon. They missed the moon but they mapped out the extension of the Van Allen belts in the radial direction. This hectic experimental activity during 1958 determined some main characteristics of the Van Allen belts rather well.

Soviet scientists also participated and contributed in important ways to these early investigations. We are glad to have with us today Konstantin Gringauz from the Space Research

Institute in Moscow, who was a leader of the early research about the radiation belts in the USSR.

The space physics group at the University of Iowa grew fast during the years after 1958 under Van Allen's leadership. During the sixties and seventies many complete satellites for magnetospheric research were built in Van Allen's department.

Van Allen and his coworkers participated in the sixties and seventies also in the very first missions to most of the planets, namely to Venus, Mars, Jupiter and Saturn. The grand tour to Jupiter, Saturn, Uranus and Neptune by the Voyagers, which reached Neptune last month, was originally planned in scientific terms by a working group of NASA with James Van Allen as chairman. Van Allen has, thus, played a key role also in the planetary exploration program of USA.

Being more than most of his fellow men at home in the whole solar system, James Van Allen is also a deeply rooted son of Iowa, this midwest agricultural state that has become internationally known for scientific achievements mainly through his and his students' work. He was born in Mount Pleasant, Iowa, on the 7th of September, 1914 as the second of four sons of Alfred and Alma Van Allen.The father was a lawyer. James Van Allen has been married to Mrs Abigail Van Allen for 44 years and they have five children, all of whom we have the great pleasure of having with us here today with their wives and husbands, and the grandchildren have also come to Stockholm.

Professor Van Allen, on behalf of the Royal Swedish Academy of Sciences I invite you to step forward and receive from the hands of his Majesty the King the 1989 Crafoord Prize.

ACTIVE EXPERIMENTS IN MAGNETOSPHERIC PHYSICS

James A. Van Allen

Department of Physics and Astronomy
The University of Iowa
Iowa City, Iowa 52242

ABSTRACT

Magnetospheric physics has been and continues to be primarily an observational science, aided and guided by theory. However, increasing emphasis is being given to active experiments for the purpose of clarifying specific phenomena in the magnetospheric plasma. The earliest global scale experiments were the production of artificial radiation belts by high altitude nuclear bomb bursts in 1958-62 and the observation of their geophysical effects. Other active experiments involve the injection of ionized clouds of exotic vapors, the decay products of radioactive sources, and beams of artificially accelerated ions and electrons at selected points in the magnetosphere. The effects of such perturbations of the natural situation as well as those of very low frequency radio waves have been observed by sophisticated diagnostic instrumentation. The *in situ* determination of the ambient electric field point-by-point in the auroral zone is one of the most important and difficult undertakings. This paper reviews the achievements of active experiments and their future promise.

1. INTRODUCTION

For several years following discovery of the radiation belts of the Earth in 1958, investigation of them and of the soon recognized and much broader range of magnetospheric phenomena was primarily observational, exploratory, and descriptive (Van Allen, 1983). Interpretation was at the level of relatively simple physical principles. My definition of a simple physical principle is one that I understand.

In 1958, which happened to be near the peak of a solar activity cycle, one light-hearted topic of our daily luncheon seminars at the University of Iowa was whether or not we would be out of work as solar activity diminished. One speculation was that strong solar activity was necessary to maintain the Earth's radiation belts and that they would fade away as solar activity diminished. The inverse speculation was that strong solar activity perturbed and tended to deplete the population of trapped particles and hence that the population would increase as solar activity diminished. Neither of these two speculations turned out to have had much merit. But they did focus attention on the diagnostic value of studying time variations and their association with external parameters of the solar wind.

Magnetospheric Physics, Edited by B. Hultqvist and C.-G. Fälthammar
Plenum Press, New York, 1990

Some of the basic physics of the magnetosphere can be inferred from a descriptive knowledge of its quasi-stationary state, the inner radiation belt of high energy protons being a prominent example. But studies of the natural dynamics of the magnetosphere, including also ones of the relationships among geomagnetic and other internal parameters of the system, have been an important part of the field during the past three decades. In some studies, a cause-and-effect relationship is relatively clear; in others, the results are bewildered by a plethora of uncontrolled parameters.

Meanwhile, major efforts have been devoted to exploratory investigation of the magnetospheric properties of the other planets (Venus, Mars, Jupiter, Mercury, Saturn, Uranus, and Neptune, in that chronological order), of comets, and of planetary satellites and rings. A certain commonality of basic principles prevails but each case is distinctively different in detail.

It has been recognized from the outset that a really thorough understanding of magnetospheric phenomena is a massively difficult objective. Many persons—now of the order of a thousand—are engaged in this undertaking. I am not competent to do justice to the immense body of current work. Indeed, I estimate that it would take me at least six months to master one month's output of published papers and conference proceedings. In other words, if I were to attempt to do this, I would fall behind by about five months every month. Worse yet, I would be unable to do any original research. Therefore, I have adopted a more modest objective for this paper.

2. EXPERIMENTS IN MAGNETOSPHERIC PHYSICS

As noted above, passive observations of natural time variations in identifiable features of the magnetosphere help delineate the physical processes involved. In addition, many suggestions have been made for artificial modifications of the system in order to examine specific effects in a controlled way. Some of these modifications have a global character; others address the micro-physics of the magnetospheric plasma within a localized region; and others are of an intermediate nature. The conceptual bases for most of these active experiments date from the free-thinking period of the late 1950s and early 1960s.

In this paper, I restrict myself to a sampling of active experiments, most of which have been reduced to practice and will likely have future applicability. My aim is to give an overview without pretense of full detail.

3. TERRELLA EXPERIMENTS

Kristian Birkeland (1908, 1913) may be said to be the father of experimental, as contrasted to observational, auroral physics. Around the turn of the century he produced energetic ($\approx 10\,\mathrm{keV}$) electron beams in a partially evacuated glass-walled chamber at the center of which was a sphere, or terrella, equipped with a current-carrying coil to simulate the magnetic Earth. He found rings of electron precipitation resembling the auroral zones and in some cases a brightness in the surrounding gas resembling a radiation belt. Many such terrella experiments have been performed subsequently and most notably by Malmfors (1946) and Block (1955). But their applicability to understanding natural aurorae and the origin of radiation belts has remained heuristic rather than conclusive because of the impossibility of scaling all of the relevant physical parameters of the Earth-system to the laboratory. Terrella experiments using electron guns in a hard vacuum have, however, been successful in faithfully demonstrating cosmic ray trajectories in a dipolar magnetic field (Malmfors, 1945).

4. NUCLEAR BOMB BURSTS AT HIGH ALTITUDE

Following a detailed, but secret, proposal by Nicholas C. Christofilos in late 1957 (Christofilos, 1966; Van Allen, 1983 and comprehensive bibliography therein), massive injections of charged particles into the Earth's magnetic field were made by thirteen high altitude nuclear bomb bursts by the United States and the Soviet Union during the period 1958–62. These are listed in Table 1 (Van Allen, 1966; Boquist and Snyder, 1967). The resulting artificial radiation belts of electrons from the decay of fission products of nine of these bursts were observed by various methods, most notably by our Explorer IV, Injun I, and Injun III satellites and by Telstar, Ariel, Explorer 12, Explorer 14, Explorer 15, and a succession of military satellites. Other observed effects of a transient nature included artificial aurorae, geomagnetic storms, synchroton radiation, very low frequency radio emissions, and massive ionospheric modifications. (Note: The term aurorae is commonly used for atmospheric luminescence produced by beams of artificially accelerated particles and by those from radioactive sources. A more proper term is pseudo-aurorae inasmuch as the natural process of accelerating auroral primaries is not simulated in such experiments.)

The number of energetic electrons from the radioactive decay of the fission products of a nuclear bomb is 3.6×10^{23} per kiloton of explosive yield. This is equal to the number of electrons from an accelerator operating at a beam current of 10 milliamperes for two months or to the number from a 1000-curie β-radioactive source in 300 years. The composite spectrum of fission-product-decay electrons is a continuous one having an upper limit of 9 MeV but 80% of the electrons have energies less than 2 MeV. In the Argus tests, a few percent of the total yield of electrons were injected into trapped orbits.

All of the bursts listed in Table 1 provided useful information. The three Argus tests, in particular, yielded a number of important advances in magnetospheric physics, viz.:

(a) The first application of the second and third adiabatic invariants to the motion of geomagnetically trapped particles (Northop and Teller, 1960);

(b) The consequent basis for McIlwain's (1961) B-L coordinate system for organizing three-dimensional data into two physically significant dimensions;

(c) Direct determination of the lifetimes of ≈ 2 MeV trapped electrons for $1.7 < L < 2.0$ (Van Allen et al., 1959);

(d) Direct determination of the geometric form of specific L = const. magnetic shells; and

(e) Upper limit estimates of some sample radial diffusion coefficients.

More broadly, the Argus tests dramatized the field of magnetospheric physics and attracted many new investigators.

The much more powerful Starfish burst (Table 1) in 1962 produced dramatic geophysical effects of a hydromagnetic character and injected a far greater number of energetic electrons 1.3×10^{25} into durably trapped orbits. Only at Jupiter has a comparable natural intensity of trapped electrons been found. The artificial population decayed to about 10% of its initial value within a year (Van Allen, 1966); but the residue of this population was detectable within the background of the natural population as late as 1968 (West and Buck, 1976).

The Soviet tests later in 1962 yielded much lower trapped particle intensities, presumably because of lower altitudes of the bursts and lesser yields.

One of the most important results of the Argus, Starfish, and Soviet bursts was a determination of the lifetimes of ≈ 2 MeV electrons in the range $1.1 < L < 3.6$ (Van Allen, 1964). At the low L values the lifetime rose rapidly (more or less consistent with estimates of quiescent atmospheric losses) but then had a maximum value of about 600 days at L=1.5. It then diminished to a value of about 30 days at L=2.0 and was approximately constant at that value

Table 1

Designation	Date of burst	Altitude (km)	Nominal explosive yield TNT equivalent	Maximun omnidirectional intensity at t = 0 $(cm^2\ sec)^{-1}$	L value of burst	Apparent mean lifetime (approx.)
Teak	1 Aug. 1958	~ 75	10 megatons	10^3	1.1	Few days
Orange	12 Aug. 1958	~ 45	10 megatons	10^3	1.1	Few days
Argus I	27 Aug. 1958	~ 200	1.4 kilotons	10^5	1.7	3 weeks
Argus II	30 Aug. 1958	~ 250	1.4 kilotons	10^5	2.1	3 weeks
Argus III	6 Sep. 1958	≳ 480	1.4 kilotons	10^6	2.0	1 month
Starfish	9 July 1962	~ 400	1.4 megatons	10^9	1.12	1.5 years
Checkmate	20 Oct. 1962	Tens of km	< 20 kilotons	–	–	–
U.S.S.R. I	22 Oct. 1962	Unknown	"submegaton"	10^7	1.9	1 month
Bluegill	26 Oct. 1962	Tens of km	< 1 megaton	–	–	–
U.S.S.R. II	28 Oct. 1962	Unknown	"submegaton"	10^7	2.0	1 month
U.S.S.R. III	1 Nov. 1962	Unknown	"megaton"	10^7	1.8	1 month
Kingfish	1 Nov. 1962	Tens of km	< 1 megaton	–	–	–
Tightrope	4 Nov. 1962	Tens of km	< 20 kilotons	–	–	–

for $2.0<L<3.6$. These lifetimes are of basic importance in the gross dynamics of the magnetosphere.

Tentative plans for additional Argus-type tests at a family of latitudes were frustrated by the 1962 Partial Test Ban Treaty, which forbids them.

The massive and long-lasting effects of the Starfish test obligate scientists to continue to remind political and military leaders of the potential consequences of any such future bursts now being considered.

5. INJECTION OF POSITRONS

Another early suggestion for tracer studies of the magnetosphere was the injection of known numbers of positrons from radioactive sources at known places and known times. The two distinctive 0.511 MeV gamma rays from a positron-electron annihilation make it possible to detect positrons in the presence of a large background of naturally present electrons, protons, and heavier ions.

A quantitative discussion of a positron-injection scheme was published by Hones (1964). In the discussion of diagnostic objectives he remarked as follows:

> "It is very important to learn what paths the high-latitude (i.e., $\lambda > 60°$) magnetic lines of force follow in the magnetosphere. These lines are distorted by the solar wind, and consequently it is not clear how, for example, particle flux measurements at 8 to 10 Earth radii (or beyond) in the equatorial plane can be related to measurements at high latitudes near the Earth. A simple positron-injection experiment would permit "connecting up" points at the earth with points, say, in the equatorial plane all along a given meridian.... It would be particularly interesting to look for precipitation, near the auroral zone, of

positrons injected at 6, 8, 10, etc. Earth radii in an attempt to determine the relation between the auroral zone and the front boundary of the magnetosphere."

The artificial injection of positrons and alpha particles from strong radioactive sources was also discussed by Christofilos (1966). Such experiments have never been conducted.

But it has been found recently that injection of detectable numbers of positrons does occur inadvertently by nuclear reactors that are flown as power supplies on large Soviet reconnaissance satellites (Rieger et al., 1989; Share et al., 1989; and Hones and Higbie, 1989). By a remarkable quirk of history, Hones is one of the participants in analyzing the positron observations by the Solar Maximum Mission (SMM) satellite. Typical L-values are, however, in the range 1.08 to 1.7; hence these observations do not yield significant new information on magnetospheric topology or dynamics.

6. INJECTION OF EXOTIC VAPORS (NEUTRAL AND IONIZABLE)

Following a suggestion of David Bates (1950) for study of the atmospheric chemistry of airglow, many rocket flights in New Mexico in the mid-1950s were conducted for the injection of sodium vapor, nitric oxide, and ethylene at altitudes of the order of 100 km (Newell, 1960). In 1958 and 1959 Soviet rockets injected a few kilograms of sodium vapor into the atmosphere at altitudes up to 430 km. The observed diffusive spreading of the neutral vapor cloud during a several minute period provided a determination of the ambient atmospheric density (Shklovskii and Kurt, 1961). The same technique was used by Shklovskii (1961) (also Shklovskii et al., 1960) in 1959 for the production and observation of artificial "comets" of sodium vapor at geocentric distances of about 120,000 km by the first and second Soviet cosmic rockets. The resonant scattering of sunlight in the sodium D lines provided an optical marker on the trajectory of the spacecraft but the cloud of neutral sodium did <u>not</u> simulate the physics of a natural comet. Shklovskii also discussed the potential superiority of (ionized) lithium.

Harrison (1962a, 1962b) advocated small nuclear bombs containing Be, Mg, Ca, Sr, or Ba to provide tracers for the direct determination of the topology of the outer reaches of the geomagnetic field. But no such experiments have been conducted.

Meanwhile, Ludwig Biermann's (1951) classical interpretation of the interaction of the solar corpuscular radiation with the ionized gaseous tails (Type I) of natural comets inspired a long series of ionospheric, magnetospheric, and interplanetary experiments by his colleagues at the Max Planck Institute in Garching/Munich and by many others. The rationale and physics for such experiments were discussed by Biermann et al. (1961) and by Biermann and Lüst (1966). They emphasized the central importance of producing clouds of rapidly ionizable gas, one of the most promising of many possibilities being high temperature vapor of barium. The physics of artificial clouds of ionized vapor is rich in plasma wave phenomena and, perhaps more importantly, provides a measurement of the electric field at a chosen point in space.

A large number of rocket payloads have been devoted to the injection of clouds of barium, lithium, cesium, strontium, and other vapors into the ionosphere, especially in the auroral zones (e.g., Föppl et al., 1965 and Haerendel et al., 1967).

A culmination of this type of investigation is represented by the recent three-spacecraft international mission AMPTE (Active Magnetosphere Particle Tracer Explorers) (Krimigis et al., 1982; Acuna et al., 1985). This mission has been executed successfully and has yielded a wide variety of results in magnetospheric plasma physics from the artificial injection of barium and lithium clouds in the magnetosheath, the upstream solar wind, and the magnetotail (Koons and Anderson, 1988). Artificial comets were produced and observed in situ as well as remotely

(Lühr et al., 1986 and related papers in *J. Geophys. Res.*, 91, No. A2, 1986). Other important observations, unrelated to the vapor releases, identified the natural ionic composition in the terrestrial ring current (e.g., Gloeckler et al., 1985) and demonstrated the importance of the ionosphere as a source of such ions during a particular magnetic storm. However, one of the central objectives of AMPTE, namely, tracing the entrance into the magnetosphere of Li^+ ions and their subsequent convection, diffusion, and energization, was not achieved (Krimigis et al., 1986). Such an experiment remains on the agenda for future consideration. It is not yet clear whether the null results bespeak a flawed concept or an inadequacy in the quantity of injected ions.

7. INJECTION OF HIGH INTENSITY PARTICLE BEAMS BY ACCELERATORS

In early explanations of geomagnetic trapping, I would start as follows: "Suppose that you had a particle accelerator at point P in the Earth's external magnetic field. Then direct the beam at an arbitrary angle a to the local magnetic vector and use Alfvén's guiding center theory to find the subsequent trajectory of the beam." Also I posed related problems: Trace the consequences of Liouville's theorem on the intensity of the beam, find the highest energy at which trapping is possible, reconcile the Alfvén theory with Størmer's first integral of the motion, etc. All of this was in the spirit of conceptual understanding.

Others decided to advocate the accelerator idea as an actual experimental technique. One example was a research paper by Hones (1965). He discussed the scheme of using chemical injection of ion tracers and showed the potential superiority of ion beams (exemplified by Ba^+ with $\approx$ 10 eV energy at currents of the order of 10 amperes) for tracing out the topology of the distant geomagnetic field, for measuring the large scale electric field, and for studying diffusion rates and possibly plasma effects. The proposed technique for observation was ground based photography using an interference filter to isolate the resonantly scattered sunlight from Ba^+ at 4554 Å, a scheme demonstrated by Shklovskii for neutral sodium. To my knowledge no ion accelerators of this nature have been flown.

The first rocket flight of an electron accelerator was conducted by Hess et al. (1971) in 1969 for the purpose of mapping magnetic lines of force by luminous streaks. Many other experiments of this nature have been conducted subsequently, as described by Winckler in a major review paper (1980) entitled "The Application of Artificial Electron Beams to Magnetospheric Research." He reviews the work of his own group at Minnesota and the work of others in the United States, France, Norway, the Soviet Union, Canada, and Japan. Typical beam energies are 20 keV, beam currents 0.5 ampere, and altitudes of injection 120 km. Winckler himself has been and continues to be a leader in this field beginning with his development of the technique in 1965.

A comprehensive variety of plasma and electromagnetic waves has been observed in the vicinity of the electron gun in the frequency range of a few kilohertz to at least 50 megahertz. These waves include "lower hybrid, whistler mode, electron cyclotron harmonics, upper hybrid, and plasma frequency emissions." Also the electron beams have generated optically observable pseudo-aurorae. The present state of this technique is illustrated by the 1988 flight of ECHO 7 (Winckler et al., 1989) from the Poker Flat Research Range near Fairbanks, Alaska. This flight employed four free-flying payloads heavily instrumented for diagnostic purposes and included a low light level TV camera in one of the payloads for optical observation at close range. An Air Force Geophysical Laboratory electron gun emitted beams of up to 40 keV in energy and 250 milliampere in current. The summit altitude of the flight was 292 km. Magnetically conjugate echoes were observed and the large-scale bounce-averaged electric field in the auroral zone was measured. Local heating of the plasma occurred and plasma waves were generated.

One of the most important challenges for the electron beam technique is the confirmation of the component of the electric field parallel to $\vec{B}$ and of one or more electrostatic double layers that now appear to be the long-sought cause of the natural acceleration of auroral particles. A recent auroral zone attempt to do this (Wilhelm et al., 1985) with pulsed electron beams has yielded suggestive evidence of the expected fast echoes but the results are not wholly persuasive, testifying to the complexities of the natural situation.

8. MEASUREMENT OF THE LOCAL ELECTRIC FIELD BY WEAK ELECTRON BEAMS

One of the most elegant space experiments to date utilizes a linear array of electron guns and a single electron beam detector on the same satellite—a technique devised by Melzner et al. (1978). If the emitted beams of electrons are directed perpendicular to the ambient magnetic field, the detection of a returning beam after one circular loop of its trajectory provides, in favorable cases, a measurement of the local electric field. Such devices were flown on the synchronous satellites GEOS-1 in 1977 and GEOS-2 in 1978–79. Typical beam currents were of the order of 0.01 microampere at an electron energy of 1.2 kilovolts. At geosynchronous altitude, a typical gyro-radius of the loop is 1 kilometer so that the flight path of the electrons is mostly remote from the parent satellite and is little affected by plasma disturbances in its near vicinity. Clear signals were detected with an electric field sensitivity of the order of 100 microvolts/meter, a value much superior to that attainable by any other in situ method. These investigations have provided marked advances in the determination of plasma convection velocities at geosynchronous altitude and the relationships of these convection patterns to parameters of the solar wind and disturbances of the geomagnetic field (Baumjohann et al., 1985; Baumjohann and Haerendel, 1985).

Attractive features of this technique are that it has a minimal effect on the ambient medium and that a measurement of the electric field at a specific point in space is obtained in a conceptually clean way. The component of $\vec{E}$ perpendicular to $\vec{B}$ is the one most readily measured but it may be possible to measure the parallel component also (Pirre, 1982).

9. VERY LOW FREQUENCY (VLF) ELECTROMAGNETIC WAVES

Very low frequency electromagnetic waves are defined to be those in the frequency range 3 to 30 kilohertz. As suitably detected, they yield sound waves within and somewhat above the range of human hearing.

Whistling atmospherics or "whistlers" were discovered by Barkhausen in 1919. In later work he and Eckersley established lightning strokes in the lower atmosphere as the cause of whistlers and discussed the apparent propagation of the waves along magnetic lines of force from one hemisphere to the other, developing the declining tone or whistle as a result of dispersion in the ambient plasma. The modern era of the subject dates from the extensive study of whistlers by Owen Storey (1953). Among the many results of his work was the inference of an electron number density of approximately 400 cm^{-3} at a height of about two Earth radii on the equator. This value was, at that time, thought to be unreasonably large but it has received massive confirmation by later indirect and direct observations in the region now called the plasmasphere (Carpenter and Smith, 1964; Helliwell, 1965).

In the hands of Helliwell and his collaborators, VLF transmitters have become an important diagnostic tool for probing the plasmasphere and the related distribution of trapped energetic particles. Ground-based transmissions have been shown to induce precipitation of energetic particles by way of wave-particle interactions, and the ducted VLF signals have been observed by satellite-borne receivers (Inan et al., 1977, 1985; Anderson, 1983). A comprehensive knowledge of the plasmasphere has been obtained by the VLF technique (Carpenter and Smith, 1964; Carpenter and Park, 1973; Chappell, 1972). In a sense, this technique is the inverse of stimulating plasma waves and VLF emissions by artificial particle beams.

A planned Soviet mission called ACTIVE will use a 5 kW VLF transmitter and a large loop antenna on a satellite in low earth orbit for the investigation of radiowave propagation and wave-particle interactions in the magnetosphere.

(During the symposium, R. Z. Sagdeev reported that the launch of ACTIVE had been accomplished successfully on 28 September 1989. The primary satellite has a loop antenna of 20 meter diameter and a pulsed transmitter operating at 9.6 ± 0.6 kHz. Alternatively, a dipole antenna of 15 meter length can radiate 25 watts in the same frequency range. Also a plasma generator can eject a plasma beam of 20 amperes at an initial velocity of 14 km s^{-1}. An accompanying Czechoslovakian subsatellite is outfitted with an extensive array of diagnostic instruments. In addition, ground based and rocket observations are planned.)

The depletion of the population of magnetospheric particles by the 60 Hz electromagnetic field from terrestrial power sources has been inferred from statistical evidence involving the "weekend" effect during which terrestrial power usage diminishes as does particle precipitation. This effect, if indeed correctly identified, is small (Anderson, 1983).

10. DEPLETION OF THE INNER RADIATION BELT BY INERT MATERIAL

An early suggestion of S. F. Singer (circa 1960, private communication) was that large "sweeper" satellites be used to deplete the population of energetic particles in the inner radiation belt and thus diminish the hazard to space flight. This suggestion has not been implemented purposefully. However, Konradi (1988) has estimated recently that the accumulation of orbiting manmade objects will, by the year 2010, have a detectable effect on the equilibrium intensity of energetic protons ($E_p \approx 55$ MeV) having L in the range 1.2 to 1.6 and mirror points in the altitude range 200 to 500 km over the South Atlantic geomagnetic anomaly.

11. PLASMA EFFECTS IN THE VICINITY OF LARGE ORBITING BODIES

The high speed flight of large orbiting bodies through the ionospheric/magnetospheric plasma produces many plasma physical effects and in itself may be considered an experiment. This subject has been advanced recently by Shawhan (1982) and his many collaborators in two successful flights of a Plasma Diagnostics Package (PDP) on space shuttles. In the first flight, the PDP was deployed by the Remote Manipulator System to explore the nearby environment of the shuttle. In the second flight, the PDP was released as a free-flying subsatellite and observations were made in various relationships to the shuttle out to distances of some 400 meters. A pulsed electron gun was also a part of the experimental regimen. The effects of the electron beam on the plasma environment as well as the effects of water dumps, jet firings, etc. were observed. The PDP was retrieved and returned to the laboratory. These experiments have yielded a substantial body of definitive knowledge on the plasma physical environment of the shuttle as well as additional insight on natural processes (e.g., Frank et al., 1989; Farrell et al., 1989; Murphy et al., 1989; Reeves et al., 1988).

12. AN ULTIMATE EXPERIMENT

One of the early international conferences on magnetospheric physics was held in Bergen, Norway in late summer 1965. During a free evening Nicholas Christofilos and I had whale-steak dinners together at a small restaurant on the waterfront. Conversation turned to how we were ever going to understand the magnetosphere. We arrived at a simple and elegant proposal. We would wrap a coil of wire around the Earth and pass the necessary electrical current through it so as to temporarily degauss the external geomagnetic field, thereby permitting the escape of all of the plasma and energetic particles. Then we would switch off the current, start with a vacuum and observe the regrowth of the system by a suitable family of instrumented satellites. Better yet, we would make artificial injections of plasma and energetic particles and observe the results. When things became too complex, we would simply repeat the degauss cycle and start again. The necessary number of ampere-turns is about 10^9. Other details were left as an exercise for the student.

This proposal has not been published previously, for obvious reasons. But I have been encouraged to mention it by President Reagan's Strategic Defense Initiative and by the 1986 report of the National Commission on Space, chaired by Thomas O. Payne. In the context of such proposals, it now seems that equipping the Earth with a Helmholtz coil might be considered a minor side job. Also degaussing the Earth may well have military value, such as disabling magnetic compasses and other effects not yet discussed openly.

Otherwise we may have to wait perhaps one million years for a natural reversal of the Earth's general magnetic field, an even less attractive possibility.

REFERENCES

Acuna, M. H., Ousley, G. W., McIntire, R. W., Bryant, D., and Paschmann, G., 1985, Editorial AMPTE-mission overview, *IEEE Trans. Geosci. Remote Sensing,* GE–23:175. (This Special Issue includes 22 other papers on specific instruments and scientific objectives.)

Anderson, R. R., 1983, Plasma waves in planetary magnetospheres, *Rev. Geophys. Space Phys.,* 21:474.

Bates, D. R., 1950, A suggestion regarding the use of rockets to vary the amount of atmospheric sodium, *J. Geophys. Res.,* 55:347.

Baumjohann, W., and Haerendel, G., 1985, Magnetospheric convection observed between 0600 and 2100 LT: solar wind and IMF dependence, *J. Geophys. Res.,* 90:6370.

Baumjohann, W., Haerendel, G., and Melzner, F., 1985, Magnetospheric convection observed between 0600 and 2100 LT: variations with K_p, *J. Geophys. Res.,* 90:393.

Biermann, L., 1951, Kometenschweife und solare korpuskularstrahlung, *Zeitschrift für Astrophysik,* 29:274.

Biermann, L., Lüst, R., Lüst, Rh., und Schmidt, H. U., 1961, Zur untersuchung des interplanetaren mediums mit hilfe künstlich eingebrachter ionenwolken, *Zeitschrift für Astrophysik,* 53:226.

Biermann, L., and Lüst, R., 1966, The interaction of the solar wind with comets (natural and artificial), *in:* "The Solar Wind" (JPL Technical Report No. 32-630), R. J. Mackin, Jr. and M. Neugebauer, eds., Pergamon Press, Elmsford, New York.

Birkeland, Kr., 1908, 1913, "The Norwegian Aurora Polaris Expedition, 1902–1903," Vol. 1, First and Second Sections, H. Aschehoug & Co., Christiania (Oslo).

Block, L., 1955, Model experiments on aurorae and magnetic storms, *Tellus,* 7:65.

Boquist, W. P., and Snyder, J. W., 1967, Conjugate auroral measurements from the 1962 U.S. high altitude nuclear test series, *in:* "Aurora and Airglow," B. M. McCormac, ed., Reinhold Pub. Corp., New York.

Carpenter, D. L., and Park, C. G., 1973, On what ionospheric workers should know about the plasmapause-plasmasphere, *Rev. Geophys. Space Phys.,* 11:133.

Carpenter, D. L., and Smith, R. L., 1964, Whistler measurements of electron density in the magnetosphere, *Rev. Geophys.,* 2:415.

Chappell, C. R., 1972, Recent satellite measurements of the morphology and dynamics of the plasmasphere, *Rev. Geophys. Space Phys.,* 10:951.

Christofilos, N. C., 1966, Sources of artificial radiation belts, "Radiation Trapped in the Earth's Magnetic Field," B. M. McCormac, ed., D. Reidel Pub. Co., Dordrecht, Holland.

Farrell, W. M., Gurnett, D. A., and Goertz, C. K., 1989, Coherent Cerenkov radiation from the Spacelab 2 electron beam, *J. Geophys. Res.,* 94:443.

Föppl, H., Haerendel, G., Loidl, J., Lüst, R., Melzner, F., Meyer, B., Neuss, H., and Rieger, E., 1965, Preliminary experiments for the study of the interplanetary medium by the release of metal vapour in the upper atmosphere, *Planet. Space Sci.,* 13:95.

Frank, L. A., Paterson, W. R., Ashour-Abdalla, M., Schriver, D., Kurth, W. S., Gurnett, D. A., Omidi, N., Banks, P. M., Rush, R. I., and Raitt, W. J., 1989, Electron velocity distributions and plasma waves associated with the injection of an electron beam into the ionosphere, *J. Geophys. Res.,* 94:6995.

Gloeckler, G., Wilken, B., Stüdemann, W., Ipavich, F. M., Hovestadt, D., Hamilton, D. C., and Kremser, G., 1985, First composition measurements of the bulk of the storm-time ring current (1-300 keV/e) with the AMPTE-CCE, *Geophys. Res. Lett.,* 12:325.

Haerendel, G., Lüst, R., and Rieger, E.,1967, Motion of artificial ion clouds in the upper atmosphere, *Planet Space Sci.,* 15:1.

Harrison, E. R., 1962a, Determination of the nature of the Earth's distant magnetic field, *Nature,* 193:359.

Harrison, E. R., 1962b, An experiment to determine the nature of the Earth's distant magnetic field, *Geophys. J., Royal Astron. Soc.,* 6:462.

Helliwell, R. A., 1965, "Whistlers and Related Ionospheric Phenomena," Stanford University Press, Stanford, CA.

Hess, W. N., Trichel, M. G., Davis, T. N., Beggs, W. C., Kraft, G. E., Strassinopoulos, E., and Maier, E.J.R., 1971, Artificial aurora experiment: Experiment and principal results, *J. Geophys. Res.,* 76:6067.

Hones, E. W., and Higbie, P. R., 1989, Distribution and detection of positrons from an orbiting nuclear reactor, *Science,* 244:448 and 1244.

Hones, E. W., Jr., 1964, On the use of positrons as tracers to study the motions of electrons trapped in the Earth's magnetosphere, *J. Geophys. Res.,* 69:182.

Hones, E. W., Jr., 1965, On the use of electrical ion sources for ion tracer experiments in the magnetosphere, Research Report, University of Iowa 65-31.

Inan, U. S., Bell, T. F., Carpenter, D. L., and Anderson, R. R., 1977, Explorer 45 and Imp 6 observations in the magnetosphere of injected waves from the Siple station VLF transmitter, *J. Geophys. Res.,* 82:1177.

Inan, U. S., Chang, H. C., Helliwell, R. A., Imhof, W. L. , Reagan, J. B., and Walt, M., 1985, Precipitation of radiation belt electrons by man-made waves: A comparison between theory and measurement, *J. Geophys. Res.,* 90:359.

Konradi, A., 1988, Effect of the orbital debris environment on the high-energy Van Allen proton belt, *Science,* 242:1283.

Koons, H. C., and Anderson, R. R., 1988, A comparison of the plasma wave spectra for the eight AMPTE chemical releases, *J. Geophys. Res.*, 93:10,016.

Krimigis, S. M., Haerendel, G., Gloeckler, G., McIntire, R. W., Shelley, E. G., Decker, R. B., Paschmann, G., Valenzuela, A., Potemra, T. A., Scarf, F. L., Brinca, A. L., and Luki, H., 1986, AMPTE lithium tracer releases in the solar wind: observations inside the magnetosphere, *J. Geophys. Res.*, 91:1339.

Krimigis, S. M., Haerendel, G., McIntire, R. W., Paschmann, G., and Bryant, D. A., 1982, The active magnetospheric particle tracer explorers (AMPTE) program, EOS, Trans. Am. Geophysical Union, 63:843.

Lühr, H., Southwood, D. J., Klocker, N., Acuna, M., Häusler, B., Dunlap, M. W., Mier-Jedrzejowicz, W.A.C., Rijnbeek, R. P., and Six, M., 1986, In situ magnetic field measurements during AMPTE solar wind Li^+ releases, *J. Geophys. Res.*, 91:1261.

Malmfors, K. G., 1945, Determination of orbits in the field of a magnetic dipole with applications to the theory of the diurnal variation of cosmic radiation, *Arkiv för Matematik, Astronomi och Fysik,* 32A(8):1.

Malmfors, K. G., 1946, Experiments on the aurorae, *Arkiv för Matematik, Astronomi och Fysik,* 34B(1):1.

McIlwain, C. E., 1961, Coordinates for mapping the distribution of magnetically trapped particles, *J. Geophys. Res.*, 66:3681.

Melzner, F., Metzner, G., and Antrack, D., 1978, The GEOS electron beam experiment S329, *Space Science Instrumentation,* 4:45.

Murphy, G. B., Reasoner, D. L., Tribble, A., D'Angelo, N., Pickett, J. S., and Kurth, W. S., 1989, The plasma wake of the shuttle orbiter, *J. Geophys. Res.*, 94:6866.

Newell, H. E., Jr., 1960, The upper atmosphere studied by rockets and satellites, *in:* "Physics of the Upper Atmosphere," J. A. Ratcliffe, ed., Academic Press, New York.

Northrop, T. G., and Teller, E., 1960, Stability of the adiabatic motion of charged particles in the Earth's field, *Phys. Rev.*, 117:215.

Pirre, M., 1982, The artificially injected charged particles as a tool for the measurement of the electric field in the magnetosphere, *in:* "Artificial Particle Beams in Space Plasma Studies," B. Grandal, ed., Plenum Press, New York.

Reeves, G. D., Banks, P. M., Neubert, T., Bush, R. I., Williamson, P. R., Fraser-Smith, A. C., Gurnett, D. A., and Raitt, W. J., 1988, VLF wave emissions by pulsed and DC electron beams in space. Spacelab 2. Observations, *J. Geophys. Res.*, 93:14,699.

Rieger, E., Vestrand, W. T., Forrest, D. J., Chupp, E. L., Kanbach, G., and Reppin, C., 1989, Man-made transients observed by the gamma-ray spectrometer on the solar maximum mission satellite, *Science,* 244:441.

Share, G. H., Kurfess, J. D., Marlow, K. W., and Messina, D. C., 1989, Geomagnetic origin for transient particle events from nuclear reactor-powered satellites, *Science,* 244:444 and 1244.

Shawhan, S. D., 1982, Description of the plasma diagnostics package (PDP) for the OSS-1 shuttle mission and JSC plasma chamber test in conjunction with the fast pulse electron gun (FPEG), *in:* "Artificial Particle Beams in Space Plasma Studies," B. Grandal, ed., Plenum Press, New York.

Shklovskii, I. S., 1961, An artificial comet as a method for optical tracking of cosmic rockets, *in:* "Artificial Earth Satellites," Vols. 3, 4, and 5, L. V. Kurnosova, ed., Plenum Press, Inc., New York.

Shklovskii, I. S., Esipov, V. F., Kurth, V. G., Moroz, V. I., and Shcheglov, P. V., 1960, An artificial comet, *Soviet Astronomy AJ,* 3:986 (Translation of *Astron. Journal of Academy of Sciences of the USSR,* Vol. 36, No. 6, Nov.–Dec. 1959).

Shklovskii, I. S., and Kurt, V. G., 1961, The determination of the density of the atmosphere at an altitude of 430 kilometers by the sodium vapor diffusion method, *in:* "Artificial Earth Satellites," Vols. 3, 4, and 5, L. V. Kurnosova, ed., Plenum Press, Inc., New York.

Storey, L.R.O., 1953, An investigation of whistling atmospherics, *Philos. Trans. R. Soc. London,* 246A:113.

Van Allen, J. A., 1964, Lifetimes of geomagnetically-trapped electrons of several MeV energy, *Nature,* 203:1006.

Van Allen, J. A., 1966, Spatial distribution and time decay of the intensities of geomagnetically trapped electrons from the high altitude nuclear burst of July 1962, *in:* "Radiation Trapped in the Earth's Magnetic Field," B. M. McCormac, ed., D. Reidel Pub. Co., Dordrecht-Holland.

Van Allen, J. A., 1983, "Origins of Magnetospheric Physics," Smithsonian Institution Press, Washington, D.C.

Van Allen, J. A., McIlwain, C. E., and Ludwig, G. H., 1959, Satellite observations of electrons artificially injected into the geomagnetic field, *J. Geophys. Res.,* 64:877.

West, H. I., Jr., and Buck, R. M., 1976, Energetic electrons in the inner belt in 1968, *Planet. Space Sci.,* 24:643.

Wilhelm, K., Bernstein, W., Kellogg, P. J., and Whalen, B. A., 1985, Fast magnetospheric echoes of energetic electron beams, *J. Geophys. Res.,* 90:491.

Winckler, J. R., 1980, The application of artificial electron beams to magnetospheric research, *Rev. Geophys. Space Phys.,* 18:659.

Winckler, J. R., Malcolm, P. R., Arnoldy, R. L., Burke, W. J., Erickson, K. N., Ernstmeyer, J., Franz, R. C., Hallinan, T. J., Kellogg, P. J., Monson, S. J., Lynch, K. A., Murphy, G., and Nemzek, R. J., 1989, ECHO 7—An electron beam experiment in the magnetosphere, *EOS, Trans. Am. Geophysical Union,* 70:657 and 666.

ACHIEVEMENTS OF MAGNETOSPHERIC RESEARCH

Bengt Hultqvist

Swedish Institute of Space Physics
Box 812
S-981 28 Kiruna
Sweden

ABSTRACT

Achievements of magnetospheric research, which are of major importance not only for the field itself but also from a basic plasma physics or general astrophysics point of view, are reviewed. Examples of such achievements are presented and discussed from the following fields: the Van Allen belts, collisionless shocks, cellular structure of the solar system, interaction at the magnetopause, electron acceleration into the atmosphere, macroscopic instabilities in magnetospheres, interaction between hot and cold magnetospheric plasmas, and wave-particle interactions.

1. INTRODUCTION

Magnetospheric physics is a creation of the space era. It became the first new scientific discipline conceived in the space age, although it was rather called space plasma physics, auroral physics or ionospheric physics in the first years, since the concept magnetosphere did not exist then. The word magnetosphere was introduced by Thomas Gold in 1959. In the three decades that this concept has existed, the research field magnetospheric physics has reached a certain degree of maturity. It is, however, still a young research field in the sense that unexpected, "surprising" results continue to constitute the most important new results from practically all new space missions. The first great surprise of magnetospheric research was the Van Allen belts, for the discovery of which Prof. Van Allen is honored during these days in Stockholm.

That the concept Magnetospheric Physics did not exist when Sputnik I was launched in 1957 does not mean that no physical concepts that are of major importance in magnetospheric physics were known before the space age. Several of them were in fact introduced in a paper published exactly half a century ago, in 1939, by Hannes Alfvén, the senior member and the pride of the Swedish space physics community. In that paper, with the title "Theory of Magnetic Storms", which was published in the proceedings of the Royal Swedish Academy of Sciences, ideas were introduced about how large-scale electric fields around the Earth accelerate auroral particles and produce currents, and these continue to influence strongly the magnetospheric research of today.

Magnetospheric Physics, Edited by B. Hultqvist and C.-G. Fälthammar
Plenum Press, New York, 1990

Still, we must conclude that before the satellite era our knowledge of magnetospheric physics was very limited and many early ideas have turned out to be wrong. This is not astonishing if we consider that magnetospheres are extremely complex systems. The complexity is so great that it is very difficult, if not impossible, to derive more than a very limited amount of information about them directly from basic principles. Only *in situ* observations can generally tell what happens and theory needs a strong guidance from experiments in order to find its right direction. A consequence of this situation is just the fact that the most important results that we have obtained from space research in magnetospheres of the planets during the last three decades have come upon us as surprises.

To learn to know and understand the magnetospheres of our solar system, of which one is the entire heliosphere, is of course an important task from a basic scientific point of view, and it is important in particular because the same basic processes as in magnetospheric physics dominate in the plasma universe, which is believed to be by far the major part of the entire universe. Magnetospheric physics is of course also important because the Earth´s magnetosphere is a part of the environment of man.

What magnetospheric research has first of all done hitherto is to provide us with an enormous amount of new knowledge and understanding of magnetospheres. We have learnt very much about their characteristics. Many of the experimental results have not yet been well interpreted in theoretical terms. We are using our own planet´s magnetosphere as a laboratory, where we carry out detailed investigations of physical phenomena of importance in many, if not all, magnetospheres. Although different magnetospheres have very different sizes and partly different characteristics, most of the important physical processes in them are the same. A main task of the research has been, and is, to identify the most important processes, i.e. the prevailing processes among the possible ones.

It is impossible to provide in the limited time available anything like a comprehensive overview of all important results of magnetospheric research. I will, therefore, limit myself to giving some examples of results from magnetospheric research which are of major importance not only for the field itself but also from a basic plasma physics or general astrophysics point of view.

2. EXAMPLES OF MAJOR ACHIEVEMENTS

2.1. The Van Allen Belts

Starting with the first great discovery of magnetospheric research, that of the Van Allen belts, an early diagram of which is shown in Fig. 1, we may today conclude that the discovery has been important first of all because it demonstrated that large fluxes of energetic particles are naturally trapped in magnetospheres in forbidden orbits, i.e. orbits which cannot be reached by particles in a static magnetosphere either from outside or from inside. Such trapped particle populations are of general importance as sources of electromagnetic radiation and of quasistatic magnetic fields. Strong radio emissions from Jupiter were among the first to be interpreted in terms of radiation from electrons trapped in a magnetic field (Warwick, 1961)

The investigations of the trapped particles in the inner magnetosphere have also made it possible to understand quantitatively the ring current around the Earth and its magnetic effects, including its sources, losses and dynamics. One of the early satellite missions that contributed most to the understanding of the ring current was Explorer 45 (S^3) in the early seventies (see e.g. Williams et al., 1969, 1973, Williams and Lyons 1974 a,b). The composition of the ring current, which is very important for the magnetic effects and also for other reasons was,

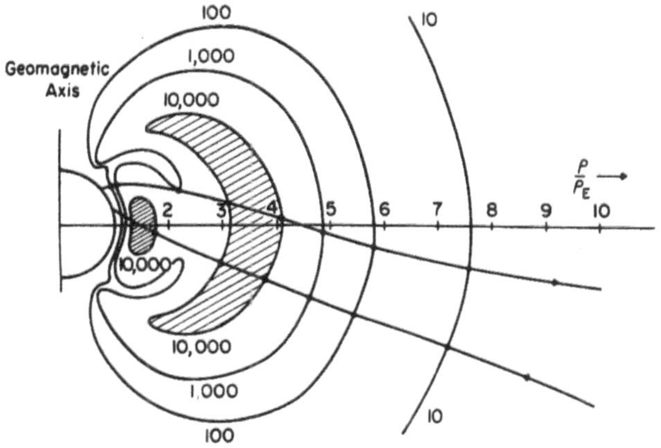

Fig. 1. A meridian cross-section of contours of equal intensity of geomagnetically trapped radiation based on data from Explorers I, III and IV and Pioneer III. The semicircle at the left represents the Earth and the two undulating curves that traverse the diagram represent the outbound (upper curve) and inbound (lower curve) trajectories of Pioneer III. The labels on the contours are counts per second of the lightly shielded Geiger-Müller tube. The linear scale is in Earth Radius (6378 km). The two cross-hatched regions of high intensity are the inner and outer Van Allen belts separated by the "slot" (after Van Allen and Frank, 1959).

however, determined only very recently from measurements with the AMPTE spacecraft by the groups at the university of Maryland, the Max-Planck-Institutes at Garching and Lindau and at Applied Physics Laboratory of the Johns Hopkins University (e.g. Hamilton et al., 1988).

The discovery of the Van Allen Belts had also another immediate effect of major importance from a general plasma physics point of view. It stimulated a very strong development of experimental and theoretical investigations of charged particle motion in magnetic and electric fields, such as found in magnetospheres. Among the main contributors were original members of the Van Allen group at Iowa, in particular Carl McIlwain. For these studies the guiding center approximation technique developed by Alfvén and published in 1940 was a basic tool.

Van Allen's discovery also gave rise to many theoretical investigations and consequent improvements of the understanding of scattering and diffusion mechanisms for the energetic particles in magnetospheres (e.g. Fälthammar, 1966; 1968). Such processes are also of basic plasma physical and astrophysical interest.

2.2. Collisionless Shocks

The bow shock in front of the Earth's magnetosphere, which was discovered by Ness et al. (1964), is a physical phenomenon which had never been seen before Explorer 33 recorded it in the early sixties. It is a collisionless shock wave. The first observations of the location of the bow shock by Ness et al. (1964) are shown in Fig. 2. In shock waves known before, collisions played a key role. In the collisionless shock waves collective plasma processes have taken the place of collisions for the scattering and thermalization of the ions and electrons. Collisionless shock waves in hot magnetized plasmas are difficult to produce and investigate in a laboratory

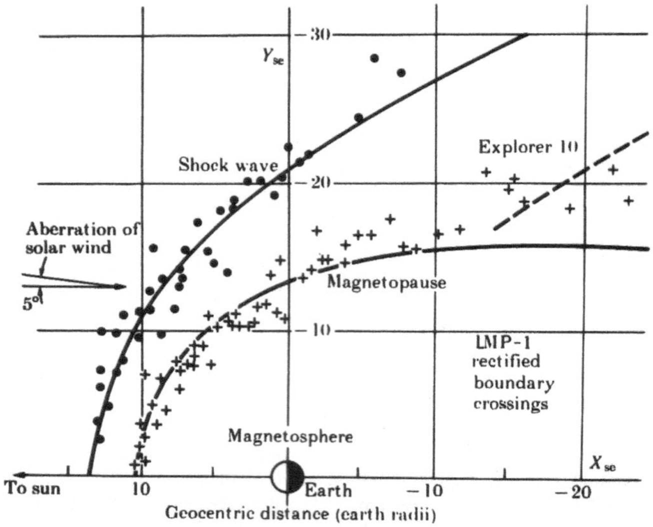

Fig. 2. The first observations of the standing shock wave in front of the Earth's magnetosphere by Ness et al. (1964).

and most of what we know about them has been learnt from space measurements. Large efforts, both experimental and theoretical, have been devoted to the investigation of such shock waves since the middle of the sixties. A representation of observational results from the Earth's bow shock is shown in Fig. 3 (after Greenstadt and Fredricks, 1979). It illustrates the complexity of the shock.

Collisionless shock waves are quite important for the acceleration of particles in the solar system. Besides the bow shocks in front of planets, many travelling shocks have been observed which are generated near the sun. There are still many aspects of the shock waves that are poorly understood and the experimental investigation of them in space will certainly go on for a long time.

2.3. Cellular Structure of the Solar System

The magnetopause, inside the bowshock, the location of which can be seen in Fig. 2, was identified by Cahill and Amazeen (1963) one year before the bowshock was discovered. Together with the thin boundary layer inside the magnetopause it separates the comparatively dense and not too hot plasma of the solar wind from the thinner and hotter plasma in the magnetosphere. Also the magnetic fields in the two regions are quite different. An important result of space plasma physics research is the demonstration that this kind of thin boundary layer between plasmas of quite different properties characterizes the solar system as a whole and not only the vicinity of the Earth. All the planets have similar boundary layers and the interplanetary magnetic field sector boundaries are another example.

The solar system thus has a kind of cellular structure and it is quite likely that this is true for the universe as a whole. The "cell walls" are not possible to observe from great distances by remote sensing techniques but only by in situ measurements. That is the reason why we have not seen them before the space age. Such a cellular structure of the plasma universe is of major astrophysical importance. Alfvén (1981) has even suggested that a consequence of the ability of plasma to concentrate differences between different populations into very thin boundary layers between them may be that a matter-antimatter symmetric universe may possibly exist, with matter and antimatter separated from each other by thin "cell walls", with too little interaction between matter and antimatter to be noticed from Earth. This is perhaps the most speculative consequence of the cellular structure but there are many other astrophysical consequences, which I shall, however, not go into here.

2.4. *Interaction at a Magnetopause*

The problem how two collisionless, high-temperature plasma populations that contain magnetic fields and move relative to each other with supersonic and super-Alfvenic velocity, interact in detail is a matter of basic importance from both a general plasma physics and an astrophysics point of view. It is a complicated and difficult matter and we are still far from anything like a complete understanding of all its aspects. But practically all existing experimental knowledge about it has been obtained from space plasma physics investigations at the boundary between the solar wind and the magnetosphere of the Earth.

The space plasma physics research has shown that the magnetosphere is not closed in the way illustrated in Fig.4, which is one of the alternative possibilities (studied first by Axford and Hines, 1961) but it is rather open, in the way proposed by Dungey (1961) and illustrated in Fig.5. How the interconnection of the magnetic fields of the solar wind and the magnetosphere - a process generally called reconnexion - really occurs is a front line item of mag-

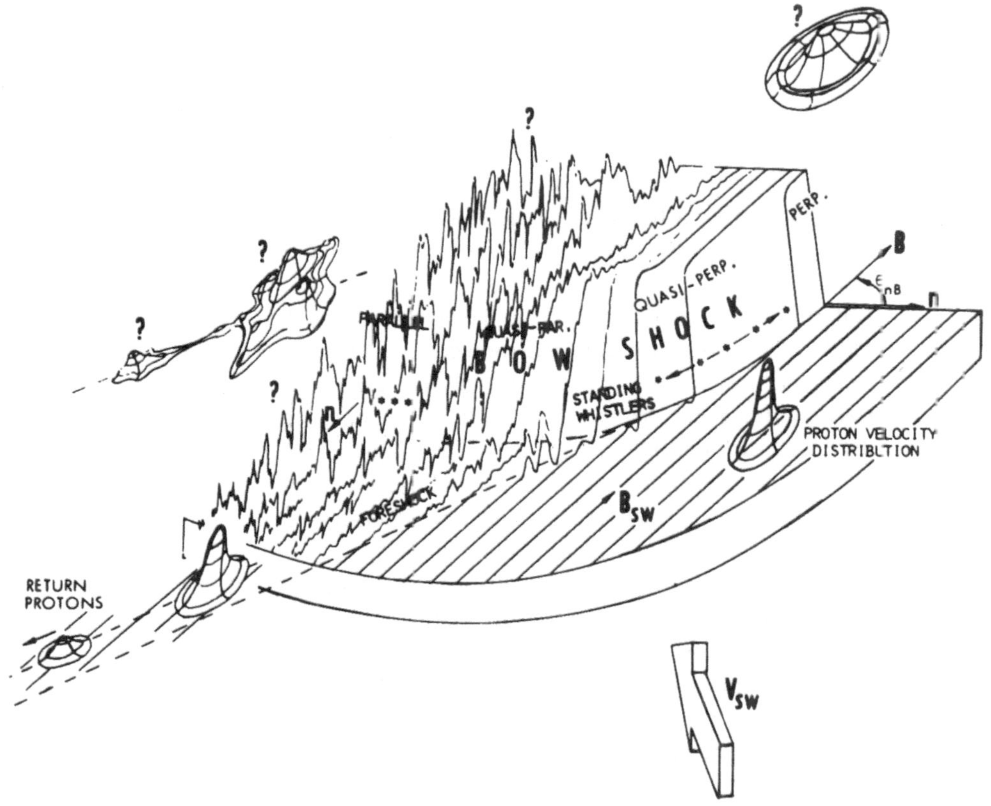

Fig. 3. Schematic representation of magnetic field profiles at different locations of the Earth's bow shock (after Greenstadt and Fredricks, 1979). Field magnitude is plotted vertically. The superposed three-dimensional sketches represent solar wind proton thermal properties as number distributions in velocity space.

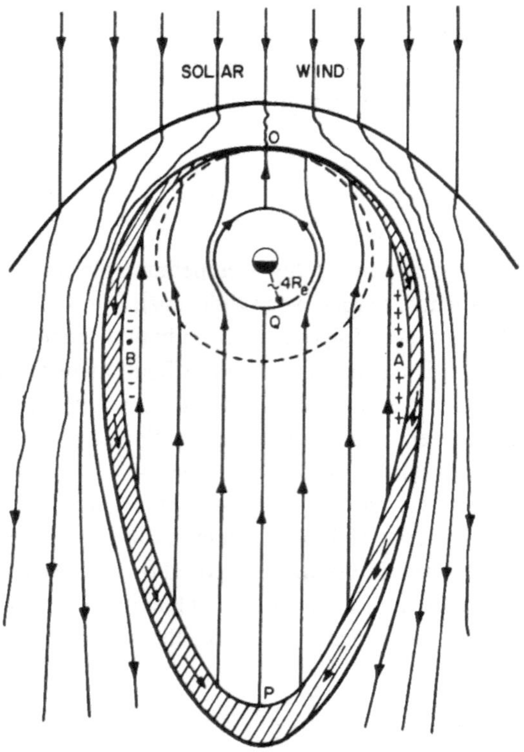

Fig. 4. Convection pattern in the equatorial plane of a closed magnetosphere, in which the geomagnetic field lines are contained, without corotation with the Earth taken into account (after Axford, 1964).

netospheric research today. The openness of the magnetic field has already been incorporated into many relevant parts of astrophysics to quite a large extent and the reconnexion process, which is still not well understood physically, has become a tool rather widely used by theoretical astrophysicists.

A result of the interaction at the magnetopause between the solar wind and the Earth is - both for a viscous kind of interaction and for reconnexion - that the electric field associated with the motion of the solar wind relative to the magnetosphere partly penetrates into the magnetosphere and forces the magnetospheric plasma to convect in ways illustrated schematically for the equatorial plane in Fig. 4. (after Axford, 1964). The achieved understanding of the general large-scale convection of the magnetospheric plasma is one of the major stepping stones for the development of our understanding of the interaction at the magnetopause as well as of many other phenomena in magnetospheric physics.

Only recently have we been able to investigate the properties of the thin magnetopause boundary layer in some detail, and we have found that the physics of the boundary layer appears to be much more complex than existing physical models suggest, with plasma components of different origin even moving in different directions with different speeds. This is illustrated by the Prognoz 7 data shown in Fig. 6 (after Lundin, 1984).

We are only at the beginning of investigating boundary layers and it is probably only at the Earth's magnetopause that we will be able to continue the detailed investigations by in situ measurements of this extremely important problem of the interaction of moving magnetized hot collisionless plasmas. That basic plasma physical and astrophysical problem will certainly stay on our agenda for a considerable time.

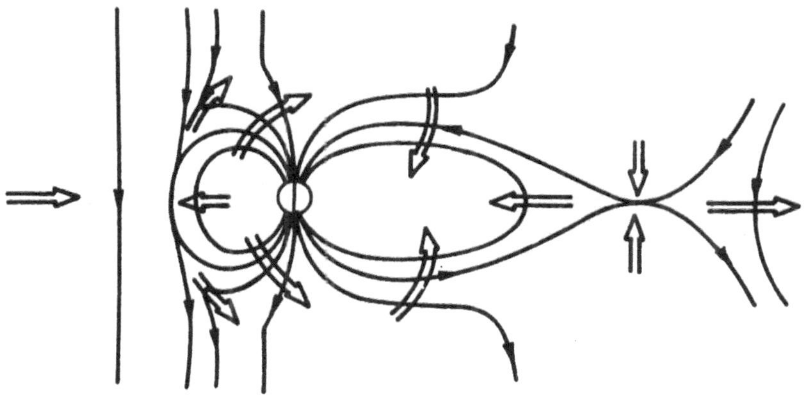

Fig. 5. The "open" magnetosphere proposed by Dungey (1961), in which the magnetic field lines of the solar wind plasma are connected with the geomagnetic field lines on the sunward side of the Earth, and they are disconnected again in the geomagnetic tail.

2.5. Electron Acceleration into the Atmosphere

One major achievement of magnetospheric research has been the demonstration that a collisionless hot magnetized plasma can sustain large potential differences in the direction of the magnetic field. That was considered as not possible according to prevailing theories only 10-20 years ago.

The first ideas about the aurora being the result of an electrical discharge around the Earth, and thus about electric fields playing a major role, date back to the eighteenth century. Alfvén seems, however, to have been the first to specify the role of the electric field in a quantitative way and to introduce an electric field component along the magnetic field lines - generally called "parallel electric fields" or, $E_{//}$ for brevity - as a major source of acceleration of the auroral energetic particles. He did this in his *Theory of magnetic Storms*, I, II, III in 1939 and 1940 mentioned earlier. Later he extended his arguments in favour of the importance of parallel electric fields in space around the Earth on the basis of experience from laboratory experiments (Alfvén, 1955, 1958) and electrostatic double layers were also proposed to be of importance in space (e.g. Alfvén, 1972).

Several leading scientists were critical of Alfvén's parallel electric field. A common argument against it was that the conductivity along the magnetic field lines in and above the ionosphere is so high that any space charges and associated parallel electric fields would be eliminated very effectively. The parallel electric field concept was thus more or less dismissed as being contrary to generally accepted theories and it was only the direct measurements in space in the 1970s, first with release experiments on sounding rockets (Haerendel et al. 1976; Wescott et al., 1976) and also with different kinds of satellite experiments that finally changed the situation completely. In the U.S.A *Review and Quadrennial Report to the IUGG* in 1979 Stern (1979) states that "in the period 1975-1978... it became increasingly evident that the condition $E_{//} = 0$ ... is often grossly violated in the magnetosphere. It would be only fair to say that the period marked a transition from general scepticism concerning the role of $E_{//}$ in the magnetosphere to the acceptance of $E_{//}$ as an essential ingredient in such phenomena as discrete auroras ... ." So it is not more than about ten years since Alfvén's 50 years old ideas became generally accepted.

Among the many experimental results that paved the way for the general acceptance of parallel electric fields, some were of special importance in breaking the resistance. To these

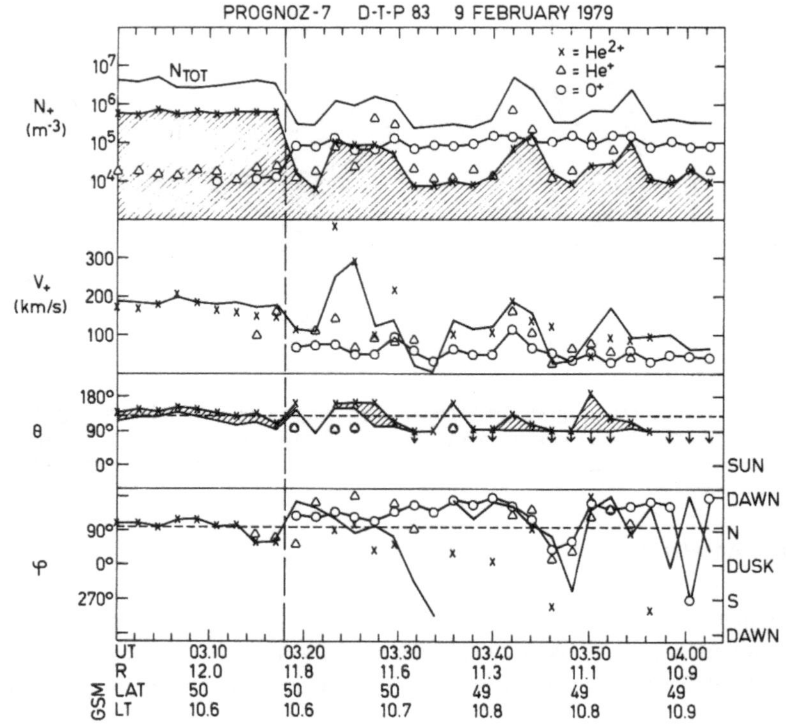

Fig. 6. Density and flow velocity of different ion species during an inbound magnetopause crossing by Prognoz 7. The magnetopause is indicated by the vertical dashed line. Panels 2-4 show the magnitude and direction of the flows of the various ion species (the solid line is for H^+) (after Lundin, 1984).

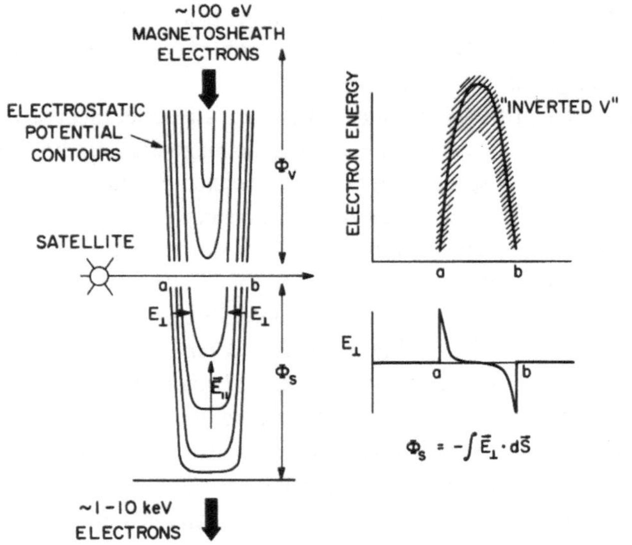

Fig. 7. Qualitative form of the electrostatic equipotential contours associated with an inverted-V electron precipitation event (after Gurnett, 1972).

belong the earlier mentioned sounding rocket result and also the discovery of the inverted V events by Frank and Ackerson (1971) which is illustrated in Fig. 7. One big stumbling block disappeared when Evans (1974) eliminated the problem that the existence of an intense low-energy tail of the observed peaked auroral electron spectra constituted (see Fig 8). The S3-3 satellite, launched in 1976, gave the final push in favour of the parallel electric field. It provided several new results in support of a parallel electric field component, one of which was the new-discovered ion beams that move out of the upper ionosphere, first reported by the Lockheed group (Shelley et al., 1976), and associated Birkeland currents, determined by Potemra and coworkers (e.g. Potemra, 1979). Another such result from S3-3 was direct electric field observations by Mozer and his group that strongly suggested the existence of electric fields with $E_{//} \neq 0$ above the auroral oval (Mozer et al., 1977). Viking has added new evidence for the existence of large potential differences along magnetic field lines (Fig. 9, after Block et al., 1987).

From the Viking satellite interesting new experimental results have been obtained also about the probable building blocks of the potential differences along the magnetic field lines, i.e. the so-called weak double layers which were observed first with the S3-3 satellite by Mozer and coworkers (see e.g. Mozer and Temerin, 1983). Boström et al. (1988) have recently found that the weak double layers appear to be a kind of rarefactive solitary waves (see Fig. 10) of dimensions of the order of a hundred meters moving upward with a speed of order tens of kilometers per second, with a parallel potential difference of the order of a volt. There may be many such solitary waves along a magnetic field line and together they may give rise to quite sizable potential differences.

The important basic question how the $E_{//}$ is produced in a hot collisionless magnetized plasma is thus a hot subject of present day magnetospheric research. The further progress on this important subject is most likely to be achieved through further experimental research in the magnetosphere of the Earth.

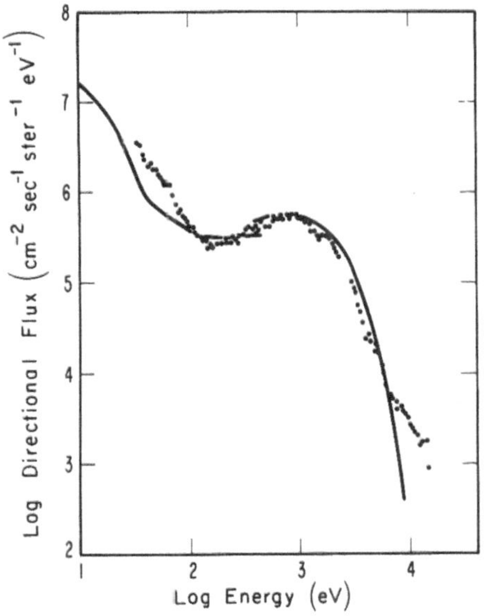

Fig. 8. Comparison of Evans' (1974) model with observations by Frank and Ackerson (1971). The electrons were assumed to have originated from an 800 eV plasma of density of 5 cm^{-3} and the field-aligned potential difference was taken to be 400 V (after Evans, 1974).

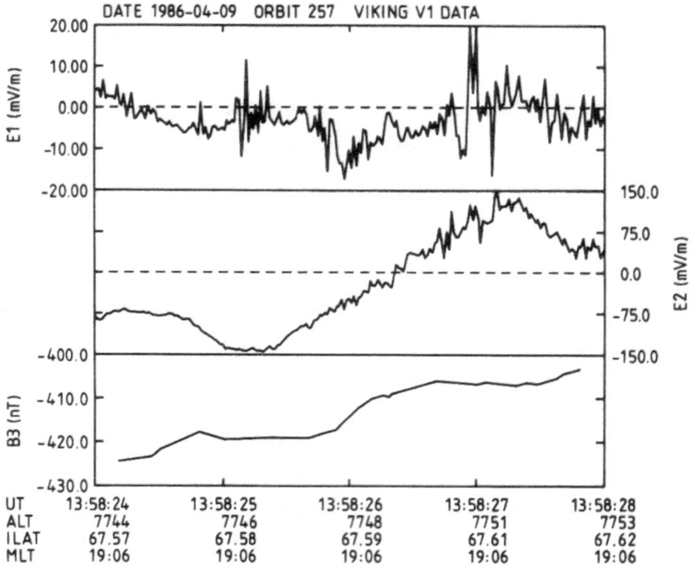

Fig. 9. Electric and magnetic field variations during a Viking pass over an evening side auroral arc. E1 is almost parallel to the local magnetic field, positive downward. E2 is perpendicular to the magnetic field and positive in the southward direction. B3 is the westward magnetic field component perpendicular to the main magnetic field. An upward parallel electric current density of about 17 μ A/m² is collocated with the maximum B3 variation, corresponding to a Birkeland current density of 50 μ A/m² in the ionosphere (after Block, 1987).

2.6. *Macroscopic Instability in Magnetospheres*

The investigations of the Earth's magnetosphere have shown that it maintains its energy and mass content within certain limits primarily by means of a macroscopic instability, which was named substorm by Chapman (see Akasofu, 1968). Substorms show up in practically all plasma and field variables in the magnetosphere and they are most probably of a similar nature as solar flares on the sun and corresponding eruptions on other stars.The interpretation of some aspects of the observational results concerning substorms is not yet quite firm, but it seems clear that reconfiguration of currents and magnetic field occurs in the tail, with the crosstail current being partly rerouted through the ionosphere. The aurora intensifies and spreads eastward and westward from the midnight region and also poleward, as determined by Akasofu and his coworkers and illustrated in the classical picture of Akasofu shown in Fig. 11. This scheme has been generally confirmed by recent satellite imaging experiments on Dynamics Explorer and Viking, although the satellite images have also shown some additional features. The entire dayside part of the auroral oval has become accessible for optical observations with the satellite-born instruments and we have found that the dayside aurora is much more interesting than we knew before. An example of a dayside aurora is shown in Fig. 12 (Plate 1) (courtesy of J.S. Murphree).

In the tail of the magnetosphere the plasma sheet has been found to become very thin during substorms some 15 to 20 earth radii from the Earth and a large plasma element, a so called plasmoid, seems to be released from the inner tail in the direction of the distant tail and out of the magnetosphere (see Fig. 13) as found by Hones and coworkers (e.g. Hones, 1979). A lot of data supporting the process illustrated in Fig. 13 has been presented, but the available data from distances beyond 25 R_e are limited in amount and not entirely conclusive. More measurements

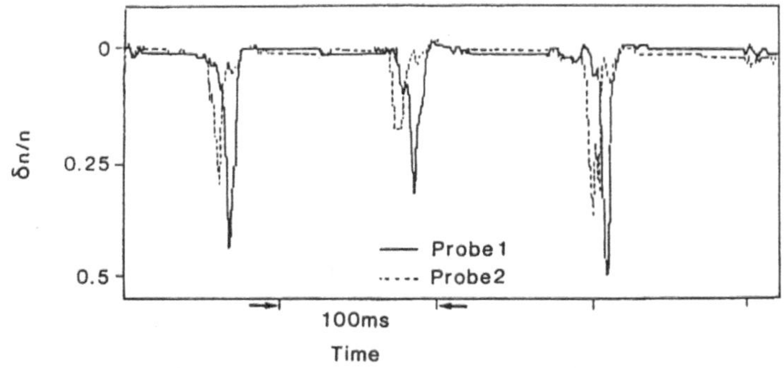

Fig. 10. Wave form data for two Δn/n probes displaying rarefactive solitary waves. The time delay between the two probe signals corresponds to an upward motion of the wave with a speed of the order of tens of kilometers per second (after Gustafsson et al., 1988).

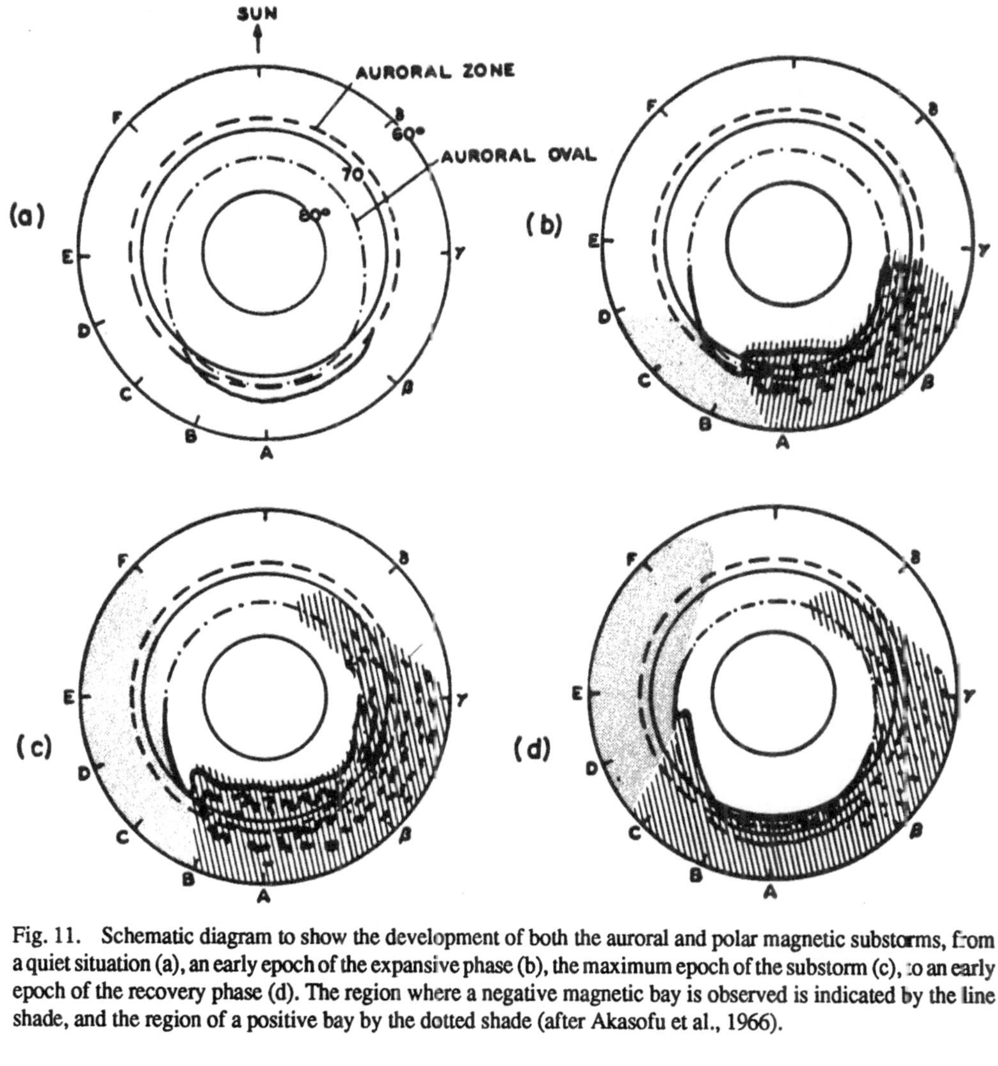

Fig. 11. Schematic diagram to show the development of both the auroral and polar magnetic substorms, from a quiet situation (a), an early epoch of the expansive phase (b), the maximum epoch of the substorm (c), to an early epoch of the recovery phase (d). The region where a negative magnetic bay is observed is indicated by the line shade, and the region of a positive bay by the dotted shade (after Akasofu et al., 1966).

are needed and will be made in the early nineties with the Soviet Interball satellites and the Japanese-US Tail satellite.

The kind of relaxation process that substorms represent is obviously of basic importance both from general plasma physics and astrophysics points of view. Again, the main laboratory for investigating these physical phenomena is the Earth's magnetosphere.

2.7. Interaction Between Hot and Cold Magnetospheric Plasmas

The energy transferred from the magnetosphere into the atmosphere through energetic particle precipitation and Joule heating ends up practically completely as heat in the upper neutral atmosphere. We have only recently learned that a large amount of magnetospheric energy is also transferred more or less exclusively to the ionized part of the upper atmosphere, i.e. the ionosphere, giving rise to heating and expulsion of ionospheric plasma into the magnetosphere.

Up until the middle of the seventies it was taken more or less for granted by the magnetospheric physics community that practically all plasma in the magnetosphere is of solar wind origin. The first measurements with ion mass spectrometers in the magnetosphere, which were carried out by the Lockheed group, gave so astonishing results of precipitating energetic O^+ ions (i.e. ions which can only be of ionospheric origin) that the experimenters awaited the confirmation of the results on a second spacecraft before publishing them (Shelley et al., 1972). The discovery in 1976 by means of the S3-3 satellite of outward flowing O^+ ions as beams above the auroral zone ionosphere (Shelley et al., 1976; Ghielmetti et al., 1978) was not quite so unexpected given the background of the earlier observations of precipitating energetic O^+ ions, but it was in some respects even more important (again obtained by the Lockheed group). The first ion mass spectrometer was sent to great altitudes on GEOS-1 in 1977 by the Bern group.

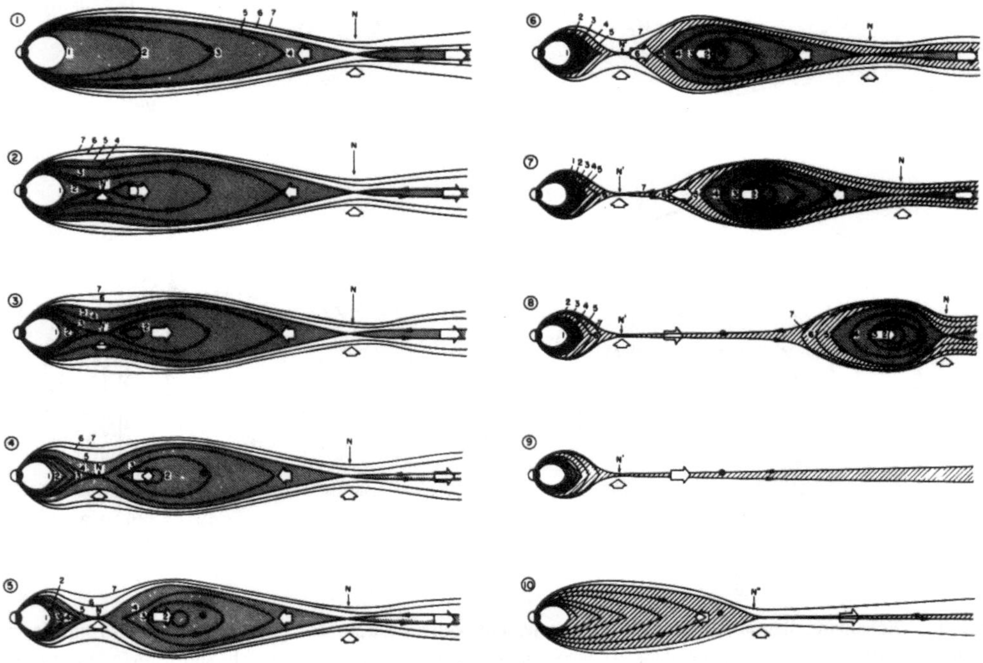

Fig. 13. Development of the magnetic field line configuration and the plasma sheet in the magnetospheric tail through a substorm (after Hones, 1978).

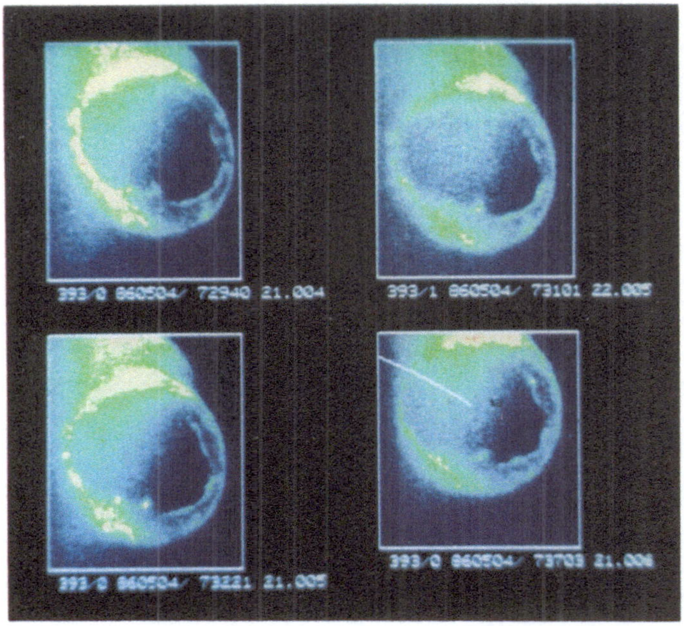

Fig. 12. A sequence of four Viking auroral images of the Northern Hemisphere auroral oval on May 4, 1986. The images illustrate a high and persistent auroral activity in the dayside oval during quiet times (K_p=2_o) with a northward directed IMF (B_z = +2 nT). Dawn is up, dusk is down (courtesy of J.S. Murphree).

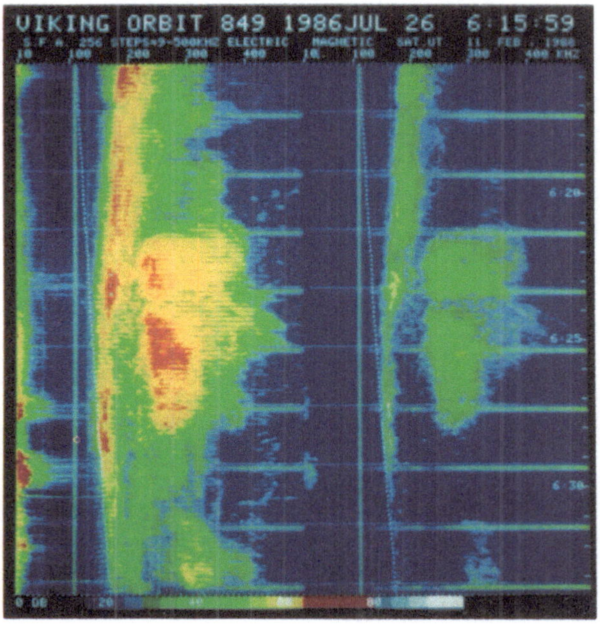

Fig. 17. Color coded spectrograms of auroral kilometric radiation (AKR) observed with Viking. Top panel shows magnetic field component and the lower panel the electric field amplitude. The dotted white lines indicate the gyrofrequency (courtesy A. Bahnsen).

Plate 1

Many research workers were then expecting that significant abundances of ionospheric ions should be found in the outer magnetosphere, but the discovery by the Bern group (Geiss et al., 1978) that the ionosphere is a source for the magnetospheric plasma of comparable importance to the solar wind came as a big surprise. Sometimes the ionospheric source dominates completely, as is shown for a magnetic storm in Fig. 14, which contains data from Prognoz 7. There is an order of magnitude higher number density of O^+ ions than of H^+ ions in the entire dayside of the magnetosphere. This was observed in a magnetic storm. Chappell and his coworkers (1987) have presented evidence, mainly from DE, for the ionosphere even being the dominant plasma source for the magnetosphere under all conditions, that means that we need no other source than the ionosphere to understand the composition of the magnetospheric plasma. This is, however, still a matter of controversy.

We have thus had in the last decade a complete revolution of our knowledge and understanding of the interaction between the ionosphere and magnetosphere in regard of ion exchange. Earlier we had greatly underestimated the effectiveness with which the magnetosphere injects energy into and extracts plasma from the ionosphere. What causes the acceleration and extraction of ionospheric plasma is still a matter of intense research. The ionospheric ions are not only accelerated along the magnetic field lines by a parallel electric field component but also perpendicularly to the field lines. The perpendicular acceleration gives rise to so called conical distributions (conics) of ion velocities at altitudes well above the height range where the acceleration takes place. Such distributions were also first reported by the Lockheed group (Sharp et al., 1977). An example of conics in an energy-pitch angle reference system is shown in Fig. 15. With Viking we have recently found that the upward moving ions correlate to almost 100% with large amplitude, i.e. hundreds of mV/m, fluctuations of the electric field, with a power density spectrum that peaks below 1 Hz (see Fig. 16). These very low dominating frequencies are thus well below any characteristic frequencies of ions in the plasma. We believe

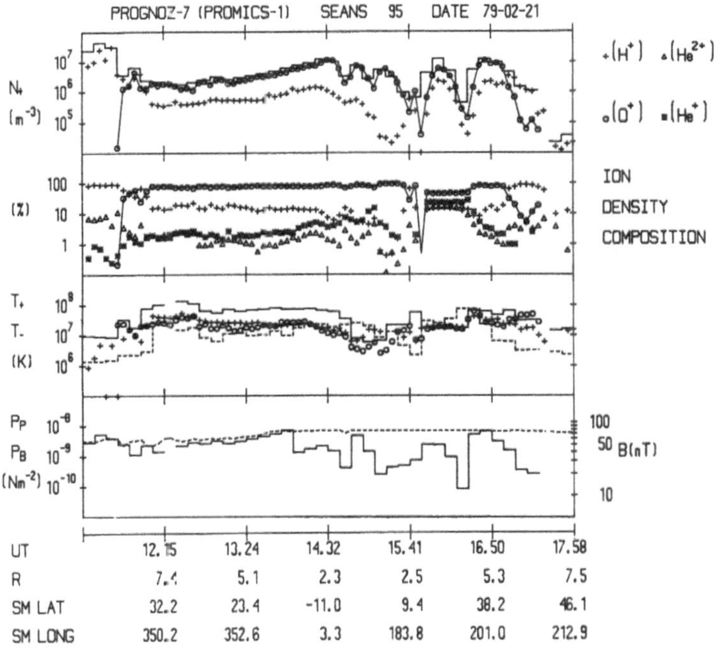

Fig. 14. An example of complete dominance of O^+ ions in the dayside magnetosphere during a magnetic storm (after Hultqvist, 1982). Solid lines show the variables based on total ion measurements.

that the ions are heated by change of ExB drift into gyro motion in the fluctuating quasistatic electric field (Lundin and Hultqvist, 1989; Hultqvist et al., 1988). How these large amplitude electric field fluctuations are produced is still unknown. The processes of acceleration and extraction of ionospheric ions are thus still only partly understood. Resonant wave-particle interactions probably also play important roles. That there is a much stronger interaction than we thought only recently is, however, undoubtable and that is of general importance for the understanding of plasma physics in general and of cosmical plasmaphysics in particular.

2.8. Wave-Particle Interactions

The magnetosphere of the Earth is an important laboratory for investigations of wave-particle interactions in hot collisionless plasma. Plasma waves play a major role in the redistribution of the magnetospheric plasma in phase space, in particular for scattering of the plasma particles into the loss cone and thereby for the transfer of magnetospheric energy into the upper atmosphere, as shown by Kennel and coworkers already in the sixties (see e.g. Kennel, 1969). Plasma waves are also important for acceleration of particles. Wave-particle interactions are responsible for the strong radio emission that several planets, including the Earth, radiate. In the case of the Earth this radiation cannot be observed from the surface because of the existence of the ionosphere. It was therefore discovered only when space probes looked back on Earth from great distances. The first evidence of radio emissions from the Earth at frequencies of the order of or less than one MHz was obtained from the Soviet Elektron 2 and 4 satellites by Benediktov et al. (1965) and there were also other indications later, but it was not until a more extensive study by Gurnett (1974) that it was clearly established that this radio emission, called auroral kilometric radiation (AKR), is generated over the auroral regions in association with bright auroral arcs. Gurnett (1974) also showed that the total power emitted

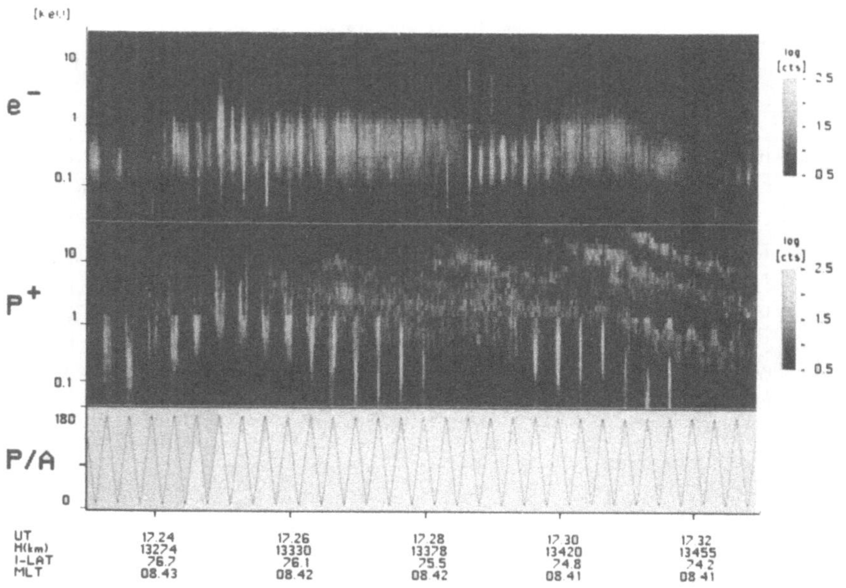

Fig. 15. Energy time/pitch angle spectrogram for ions and electrons obtained in the cleft region of the dayside auroral oval (Viking orbit 186). Each spectrogram displays counts accumulated versus energy in 32 energy steps within 0.6 s, corresponding to a pitch angle resolution of 5°. The lower panel shows pitch angle versus time. 0° corresponds to downcoming particles.

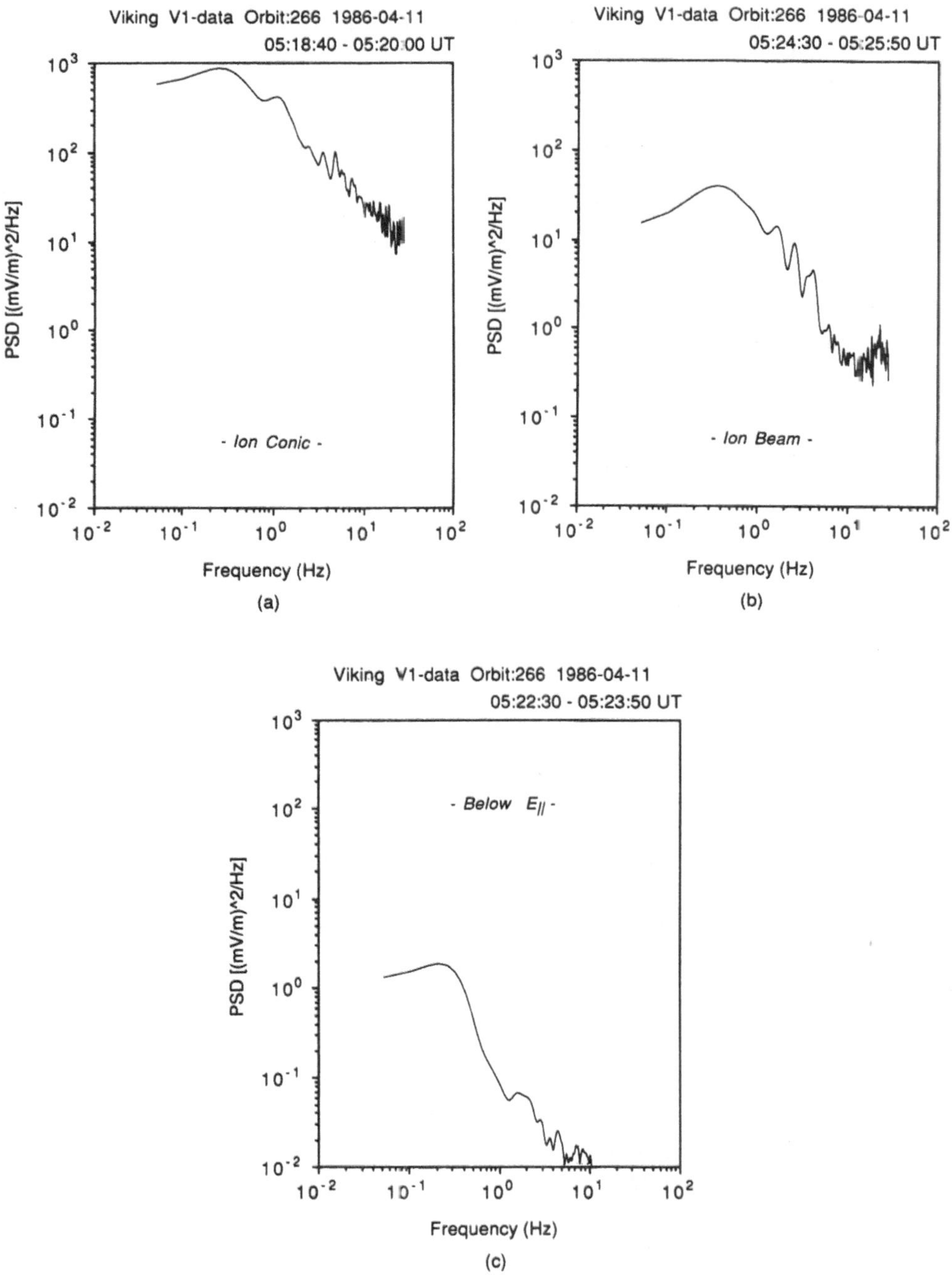

Fig. 16. Viking measurements of power density as function of frequency for electric field fluctuations observed together with ion conic distribution (a), ion beams (b) and with neither conic nor beam distribution (c) (courtesy Block and coworkers).

in the form of AKR amounts to up to 10^9 W. This is by far the most powerful radio emission in the Earth's magnetosphere and it is comparable in many respects to the strong emissions from Jupiter and Saturn. An example of observations with Viking is shown in Fig. 17 (Plate 1) (see also Bahnsen et al., 1987, 1989).Very high intensities are produced not only of propagating radio waves but also of nonpropagating electrostatic waves at much lower frequencies (Pottelette et al., 1988).

AKR is also interesting from the point of view of basic plasma physics because in order to interpret the observations a generation theory has been derived in which the relativistic factor $\sqrt{1 - v^2/c^2}$ plays a major role, in spite of the fact that all particle energies involved are clearly subrelativistic. I believe - although I am not quite sure - that this is the first time one has found such strong "relativistic" effects at subrelativistic energies.

The magnetosphere has in the last decade contributed a lot of new information about many other modes of plasma waves than those mentioned above. A lot has been learnt about microinstabilities in collisionless hot plasma. Among these many results I will here mention only one example, namely that the first direct quantitative comparison of experimental data on electrostatic wave generation in a collisionless hot plasma having a loss cone electron distribution with the kinetic theory of wave generation by such an electron distribution was made by means of measurement results obtained in space with the GEOS 1 satellite by Rönnmark et al. (1978, 1981).

3. CONCLUDING REMARKS

We are then at the end of my list of examples of basic scientific achievements based on space physics research. The selection of subjects in the list certainly reflects personal interests to some extent. If anyone thinks there has been an overrepresentation of Swedish contributions in my talk, I hope you will forgive me. There are many more examples which I have not mentioned because of lack of time and possibly of insight. I have not included any results from other planets. That has, however, been on purpose, because Tom Krimigis will next report on the very latest phase of one of the greatest, most fascinating and most successful research projects of all categories, the grand tour to the outer planets by the Voyager spacecraft and the recent encounter on the 25th of August, with Neptune, the presently most distant of the planets in the solar system.

REFERENCES

Akasofu, S.-I., C.-I. Meng, and D.S. Kimball, 1963, Dynamics of the aurora, 6, Formation of patches and their eastward motion, *J. Atmos. Terr. Phys.*, 28, 505.

Akasofu, S.-I., 1968, Polar and Magnetospheric Substorms, Preface, D. Reidel Publ. Comp. Dordrecht.

Alfvén. H., 1939, 1940, Theory of Magnetic storms, I, *Kungl. Svenska Vetenskapsakademiens Handlingar (3),* 18, No 3; II, III, ibid 18. No 9.

Alfvén, H., 1940, On the motion of a charged particle in a magnetic field, *Ark. f. mat., astron. och fysik,* 27A, No. 22.

Alfvén, H., 1955, On the electric field theory of magnetic storms and aurorae, *Tellus,* 7, 50.

Alfvén, H., 1958, On the theory of magnetic storms and aurorae, *Tellus,* 10, 104.

Alfvén, H., 1972, Relations between cosmic and laboratory plasma physics, in Cosmic Plasma Physics, p. 1, K. Schindler, ed., Plenum Press, New York.

Alfvén, H., 1981, Cosmic Plasma, D. Reidel Publ. Co., Dordrecht.

Axford, W.I., and C.O. Hines, 1961, A unifying theory of high-latitude geophysical phenomena and geomagnetic storms, *Canad. J. Phys.*, 39, 1433.

Axford, W.I., 1964, Viscous interaction between the solar wind and the Earth's magnetosphere, *Planet. Space Sci.*, 12, 45.

Bahnsen, A., M. Jespersen, E. Ungstrup, and I.B. Iversen, 1987, Auroral hiss and kilometric radiation measured from the Viking satellite, *Geophys. Res. Lett.*, 14, 471.

Bahnsen, A., B.M. Pedersen, M. Jespersen, E. Ungstrup, L. Eliasson, J.S. Murphree, R.D. Elphinstone, L. Blomberg, G. Holmgren, and L.J. Zanetti, 1989, Viking observations at the source region of auroral kilometric radiation, *J. Geophys. Res.*, 94, 6643.

Benediktov, E.A., G.G. Getmantsev, Yu.A. Sazonov, and A.F. Tarasov, 1965, Preliminary results of measurements of the intensity of distributed extra-terrestrial radio-frequency emission at 725 and 1525 kHz frequencies by the satellite Electron-2, *Kosm. Issled.*, 3, 614.

Block, L.P., C.-G. Fälthammar, P.-A. Lindqvist, G.T. Marklund, F.S. Mozer, A. Pedersen, T.A. Potemra, and L.J. Zanetti, 1987, Electric field measurements on Viking: First results, *Geophys. Res. Lett.*, 14, 435.

Boström R., G. Gustafsson, B. Holback, G. Holmgren, H. Koskinen, and P. Kintner, 1988, Characteristics of solitary waves and weak double layers in the magnetospheric plasma, *Phys. Rev. Lett.*, 61, 82.

Cahill L.J., and P.G. Amazeen, 1963, The boundary of the geomagnetic field, *J. Geophys. Res.*, 68, 1835.

Chappell, C.R., T.E. Moore, and J.H. Waite, Jr., 1987, The ionosphere as a fully adequate source of plasma for the Earth's magnetosphere, *J. Geophys. Res.*, 92, 5896.

Dungey, J.W., 1961, Interplanetary magnetic field and the auroral zones, *Phys. Rev. Lett.*, 6, 47.

Evans, D.S., 1974, Precipitating electron fluxes formed by a magnetic field aligned potential difference, *J. Geophys. Res.*, 79, 2853.

Fälthammar, C.-G., 1966, On the transport of trapped particles in the outer magnetosphere, *J. Geophys. Res.*, 71, 1487.

Fälthammar, C.-G., 1968, Radial diffusion by violation of the third adiabatic invariant, in Earth's Particles and Fields, p. 157, B.M. McCormac, ed., Reinhold, New York.

Frank, L.A., and K.L. Ackerson, 1971, Observations of charged particle precipitation into the auroral zone, *J. Geophys. Res.*, 76, 3612.

Geiss, J., H. Balsiger, P. Eberhardt, H.P. Walker. L. Weber, D.T. Young, and H. Rosenbauer, 1978, Dynamics of magnetospheric ion composition as observed by the GEOS mass spectrometer, *Space Sci. Rev.*, 22, 537.

Ghielmetti, A.G., R.G. Johnson, R.D. Sharp, and E.G. Shelley, 1978, The latitudinal, diurnal, and longitudinal distributions on upward flowing energetic ions of ionospheric origin, *Geophys. Res., Lett.*, 5, 59.

Gold, T., 1959, Motions in the magnetosphere of the Earth, *J. Geophys. Res.*, 64, 1219.

Greenstadt E.W., and R.W. Fredricks, 1979, Shock systems in collisionless space plasma, in Solar System Plasma Physics, Vol, III, p. 3, L.J. Lanzerotti, C.F. Kennel, and E.N. Parker, eds., North-Holland Publ., Co., Amsterdam.

Gurnett, D.A., 1972, Electric field and plasma observations in the magnetosphere, in Critical Problems of Magnetospheric Physics. p. 128, E.R. Dyer, ed., Nat. Acad. of Sci., Washington, DC.

Gurnett, D.A:, 1974, The Earth as a radio source: Terrestrial kilometric radiation, *J. Geophys. Res.*, 790, 4227.

Gustafsson, G., R. Boström, B. Holback, G. Holmgren, and H. Koskinen, 1988, First Viking results: Low frequency wave measurements, Presented at 6th SCOSTEP STP Sympos., Toulouse, 1986, *Physica Scripta,* 37, 475.

Haerendel G., E. Rieger, A. Valenzuela, H. Föpple, H.C. Stenbœck-Nielsen, and E.M. Wescott, 1976, in European Programmes on Sounding Rocket and Balloon Research in the Auroral Zone, ESA Ref. SP 115.

Hamilton, D.C., G. GLoeckler, F.M. Ipavich, W. Stüdemann, B. Wilken, and G. Kremser, 1988, Ring current development during the great geomagnetic storm of February 1986, *J. Geophys. Res.,* 93, 14343.

Hones, E.W., Jr., 1979, Plasma flow in the magnetotail and its implications for substorm theories, in Dynamics of the Magnetosphere by S.-I. Akasofu, ed., D. Reidel Publish, Co. Dordrecht.

Hultqvist, B., 1982, Recent progress in the understanding of the ion composition in the magnetosphere and some major question marks, *Rev. Geophys. Space Phys.,* 20, 589.

Hultqvist, B., R. Lundin, K. Stasiewicz, L. Block, P.A. Lindquist, G. Gustafsson, H. Koskinen, A. Bahnsen, T.A. Potemra, and L.J. Zanetti, 1988, Simultaneous observations of upward moving field-aligned energetic electrons and ions on auroral zone field lines, *J. Geophys. Res.,* 93, 9765.

Kennel, C.F., 1969, Consequences of a magnetospheric plasma, *Rev. Geophys.,* 7, 379.

Lundin, R., 1984, Solar wind energy transfer regions inside the dayside magnetopause-II, Evidence for a MHD generator process, *Planet. Space Sci.,* 32, 7571.

Lundin, R., and B. Hultqvist, 1989, Ionospheric plasma escape by high-altitude electric fields: Magnetic moment pumping, *J. Geophys. Res.,* 94, 6665.

Mozer, F.S., C.W. Carlson, M.K. Hudson,m R.B. Torbert, B. Parady, J. Yatteau, and M.C. Kelley, 1977, Observations of paired electrostatic shocks in the polar magnetosphere, *Phys. Rev. Lett.,* 38, 292.

Mozer, F.S., and M. Temerin, 1983, Solitary waves and double layers as the sources of parallel electric fields in the auroral acceleration region, in High Latitude Space Plasma Physics, p. 453, B. Hultqvist, and T. Hagfors, eds., Plenum Press, New York.

Ness., N.F., C.S. Scearce, and J.B. Seek., 1964, Initial results of the Imp 1 magnetic field experiment, *J. Geophys. Res.,* 69, 3531.

Potemra, T.A., 1979, Current systems in the Earth's magnetosphere, *Rev. Geophys. Space Phys.,* 17, 640.

Pottelette, R., M. Malingre, A. Bahnsen, L. Eliasson, K. Stasiewicz, R.E. Erlandson, and G. Marklund, 1988, Viking observations of bursts of intense broadband noise in the source regions of auroral kilometric radiation, *Ann. Geophys.,* 6, 573.

Rönnmark, K., H. Borg, P.J. Christiansen, M.P. Gough, and D. Jones, 1978, Banded electron cyclotron harmonic instability - A first comparison of theory and experiment, *Space Sci. Rev.,* 22, 401.

Rönnmark, K., and P.J. Christiansen, 1981, Dayside electron cyclotron harmonic emissions, *Nature,* 294, 335.

Sharp, R.D., R.G. Johnson, and E.G. Shelley, 1977, Observations of an ionospheric acceleration mechanism producing energetic (keV) ions primarily normal to the geomagnetic direction, *J. Geophys. Res.,* 82, 3324.

Shelley, E.G., R.G. Johnson, and R.D. Sharp, 1972, Satellite observations of energetic heavy ions during a geomagnetic storm, *J. Geophys. Res.,* 77, 6104.

Shelley, E.G., R.D. Sharp, and R.G. Johnson., 1976, Satellite observations of an ionospheric acceleration mechanism, *Geophys. Res. Lett.,* 3, 654.

Stern, D.P., 1979, Electric fields in the Earth's magnetosphere, *Rev. Geophys. Space Phys.*, 17, 626.

Van Allen, J.A., and L.A. Frank, 1959, Radiation around the Earth to a radial distance of 107400 kilometers, *Nature,* 203, 1006.

Warwick, J.W., 1961, Dynamic spectra of Jupiter's decametric emission, *Astrophys. J.*, 137, 41.

Wescott, E.W., H.C. Stenbæck-Nielsen, T.J. Hallinan, T.N. Davis, and H.M. Peek, 1976, The skylab barium plasma injection experiments. 2. Evidence for a double layer, *J. Geophys. Res.*, 81, 4495.

Williams, D.J., R.A. Hoffman, and G.W. Langanecker, 1969, The small scientific satellite (S^3) program and its first payload, *ISEEE Trans. on Nuclear Sci.*, NS-16, No. 1, 322.

Williams, D.J., T.A. Fritz, and A. Konradi, 1973, Observations of proton spectra ($1.0 \leq E_p \leq 300$ keV) and fluxes at the plasmapause, *J. Geophys. Res.*, 78, 4751.

Williams, D.J., and L.R. Lyons, 1974a, The proton ring current and its interaction with the plasmapause; Storm recovery phase, *J. Geophys. Res.*, 79, 4195.

Williams, D.J., and L.R. Lyons, 1974b, Further aspects of the proton ring current interaction with the plasmapause: main and recovery phases, *J. Geophys. Res.*, 79, 4791.

THE ENCOUNTER OF VOYAGER 2 WITH NEPTUNE'S MAGNETOSPHERE

S. M. Krimigis

Applied Physics Laboratory
The Johns Hopkins University
Johns Hopkins Road
Laurel, Maryland 20707-6099
U.S.A.

ABSTRACT

The particles and fields complement of instruments on the Voyager 2 spacecraft performed a comprehensive set of measurements during the encounter with the Neptune system on August 24 through 28, 1989. These included measurements of the magnetic field, plasma (10 ev $\leq$ E $\leq$ 6 keV), energetic and high energy particles (22 keV $\leq$ E $\leq$ 5 MeV), plasma waves (10 Hz to 50 kHz) and radio emissions (~ 20 to ~ 1300 kHz); additional information relating to UV emissions was provided by the ultraviolet spectrometer. The preliminary results of these measurements are reviewed in this paper and may be summarized as follows:
(a) The planetary magnetic field outside ~ 4 R_N may be described by an offset (~ 0.55 R_N), tilted (47°), dipole of moment 0.133 Gauss-R_N^3; inside that distance the field is dominated by higher order terms.
(b) Plasma densities were generally low (~ 10^{-3} cm^{-3}), except at magnetic equatorial crossings when densities up to ~ 1 cm^{-3} were seen; best fits (but not unique) to corotating Maxwellians are obtained for two components, H^+ and N^+.
(c) Energetic ions ($\geq$ 28 keV) and electrons ($\geq$ 22 keV) were seen throughout the magnetosphere, while higher energy ($\geq$ 200 keV) particles (equivalent to Van Allen belts at Earth) were confined inside the orbital radius of Triton. Spectrally soft electron and ion enhancements during spacecraft passage through the polar cap region suggest possible auroral precipitation.
(d) A variety of plasma wave emissions were seen, including chorus, hiss, electron cyclotron waves, and upper hybrid resonance in the inner magnetosphere.
(e) Radio wave bursts in the range 100 to 1300 kHz, narrowband and strongly polarized, have enabled determination of the planetary rotation period as 16.11 ± 0.05 hours; smooth emissions in the range 20 to 865 kHz were also observed.
(f) Weak auroral emissions in H Lyman β (1025Å) have been tentatively identified on the nightside of Neptune with total radiated power of 5 x 10^7 W. The magnetic field orientation with respect to the rotation axis was such that the spacecraft entered a "pole-on" magnetosphere, fortuitously sampling the cusp region of an outer planet for the first time. The measured flux of soft electrons and ions over the polar region of ~ 5 x 10^{-4} erg/cm^2 sec results in an estimated power input of ~ 3 x 10^7 W, i.e. substantially less than that at other planets. Possible reasons for these differences are discussed.

Magnetospheric Physics, Edited by B. Hultqvist and C.-G. Fälthammar
Plenum Press, New York, 1990

1. INTRODUCTION

The recent encounter of the magnetosphere of Neptune by the Voyager 2 spacecraft brings to a close an era of planetary investigations that began in the second part of the twentieth century with the launch of the first earth-orbiting satellite. Although it is probably a coincidence that the Royal Swedish Academy of Sciences decided to honor Professor James A. Van Allen this year, it is nevertheless quite appropriate that this honor should coincide with the completion of this historic period of robotic exploration of the solar system. As we all know, it was James A. Van Allen who started this era back in the late 50's with the discovery of the Earth's radiation belts that bear his name. There is, of course, another planet yet to be explored, i.e. Pluto, but given the technical means that we now have at our disposal, such an investigation will not take place until early in the 21st century.

It is a great honor for me to have been invited to present the latest findings of the Voyager spacecraft from the investigation of the magnetosphere of Neptune, for the fact that I stand on this lectern today I owe to a large extent to James A. Van Allen. As a beginning graduate student at the University of Iowa in 1961, my interest and imagination for the then brand new field of space physics was kindled by Dr. Van Allen and he offered me the opportunities to work and grow in this fascinating field of science. For his teaching and inspiration and the standards that he set for his students, I will always be grateful.

The Voyager spacecraft, like a modern-day Columbus, have set the standard for exploration in an era when technological break-through are not uncommon. Figure 1 is a schematic of the Voyager spacecraft showing the location of the eleven investigations and of spacecraft subsystems. There are five experiments that were principally intended for investigation of the plasmas, magnetic field, and wave environment of the magnetospheres of the outer planets. The magnetometer (MAG) mounted on a 13-m boom, the plasma wave (PWS) and planetary radio astronomy (PRA) instruments, both connected to two antennas located on either side of the RTG (Radioisotope Thermoelectric Generator) on the lower part of the spacecraft; the particle instruments, located on the science boom, are the plasma science (PLS), the low energy charged particles (LECP) and the Cosmic Ray experiments (CRS). Important informa-

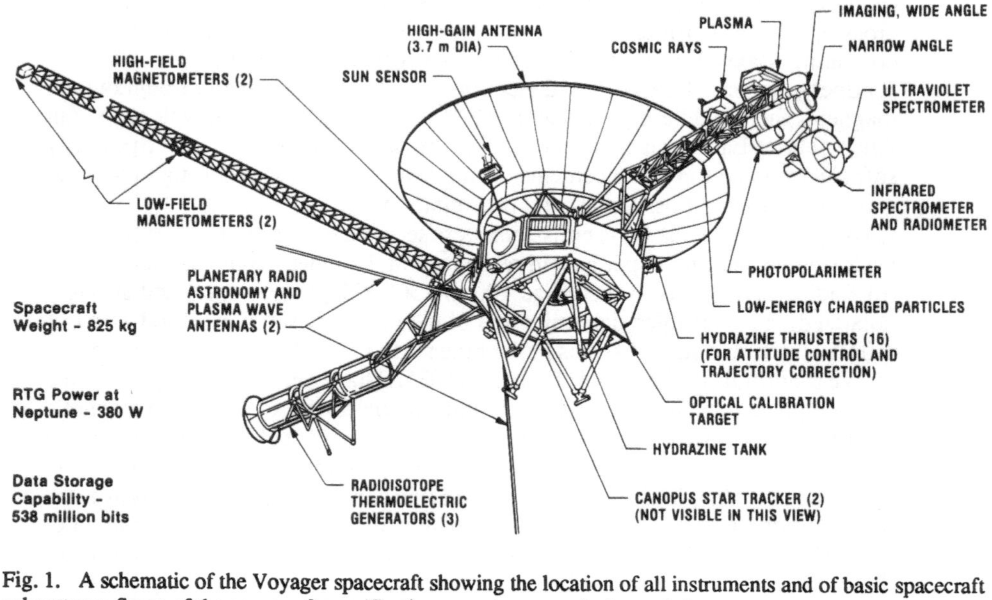

Fig. 1. A schematic of the Voyager spacecraft showing the location of all instruments and of basic spacecraft subsystems. Some of the spacecraft specifications are shown on the lower left corner.

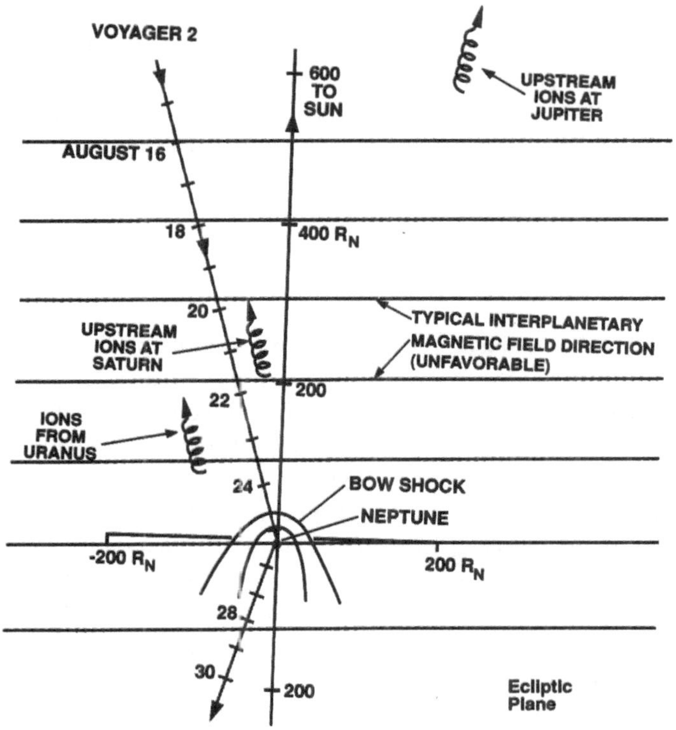

Fig. 2. Magnetosphere ions upstreams of planetary bow shocks. A projection of the Voyager 2 trajectory in the vicinity of Neptune in solar ecliptic coordinates. The interplanetary magnetic field is shown in the ideal spiral direction. The bow shock and magnetopause are drawn at their approximate locations. The spirals with the arrows indicate the onset of upstream ion detection at Saturn and Uranus along the trajectory; the upstream ions at Jupiter were actually observed at the distance of ~ 103 R_J from that planet.

tion on the aurora has been provided by the ultraviolet spectrometer (UVS), located on the scan platform on the upper right of the spacecraft. The information rate for all the non-imaging science was 3.6 kilobits per second, i.e. substantially higher than many spacecraft have had in earth orbit, especially at the time of the launch of the Voyagers in 1977. The performance of the Voyager 2 spacecraft and all its subsystems during the Neptune encounter was essentially flawless.

The encounter geometry of interest from the standpoint of the magnetosphere is shown in Figure 2. The spacecraft approached the planet at ~ 1300 local time, the trajectory being nearly perpendicular to the ideal spiral direction of the interplanetary magnetic field (IMF). Closest approach was over the north pole at an altitude of ~ 4,900 km, with the spacecraft crossing the ring plane both on the inbound and outbound legs and eventually exiting the magnetosphere at high south ecliptic latitude. The bow shock and magnetopause sketched in the figure, although drawn prior to encounter, are approximately in the correct positions as determined subsequently by the Voyager measurements. The upstream geometry is obviously unfavorable for the detection of ions emanating from the vicinity of the planetary bow shock, unless the interplanetary magnetic field turns to the radial direction due to some perturbation in the flow of the solar wind. Such, however, was not the case prior to entry of the Voyager spacecraft into the magnetosphere (Ness et al., 1989) and no upstream ions were detected by

Voyager, the first planetary encounter where this was so. The appearance of energetic ions at varying distances from the planetary bow shock in previous encounters are indicated by the spirals for the cases of Jupiter, Saturn, and Uranus; the Jupiter ions actually were first detected at a distance of over 1000 planetary radii from that planet (Krimigis et al., 1979). Most, but not all, of the data that will be discussed in the present report can be found in the special issue of the journal Science, Volume 246, December 15, 1989.

2. OBSERVATIONS

Interplanetary Environment. Since early 1989, the interplanetary environment in the vicinity of Voyager had been exceedingly disturbed. The large solar flares of early March 1989 produced strong shocks that were observed at Earth and propagated to the outer solar system where they were detected by both Voyager 1 and 2 ~ 90 days later. Figure 3 shows Voyager 2 measurements beginning on 9 June through the Neptune encounter and extending to the middle of September. Here the background has been subtracted in order to enhance the foreground signal in the low energy detector. The spectrogram on the top panel shows the intensity profile from ~ 30 keV to ~ 4 MeV, while the lower three panels display specific energy channels in the indicated energy intervals, ranging from ~ 40 keV to > 3 MeV. It is evident that there was significant activity in the interplanetary medium throughout the entire period with an approximate 26-day variation. The Neptune encounter occurred on days 236 through 239 at a time when the interplanetary activity was close to its minimum and beginning to increase again at the onset of yet another 26-day cycle. Thus, despite the presence of significant variability throughout this period, it is probable that the Neptune magnetosphere was relatively undisturbed when Voyager performed its measurements.

The first evidence that Neptune possesses a magnetic field (and associated radiation belts) came from the detection of radio bursts observed approximately five days prior to closest approach. These bursts ranged in frequency from ~ 100 to 300 kHz, were narrow-band and strongly polarized. Figure 4 shows a preliminary plot of the occurrence of these bursts related to the Neptune rotation period which was deduced to be 16h 03m ± 4m, compared to the pre-

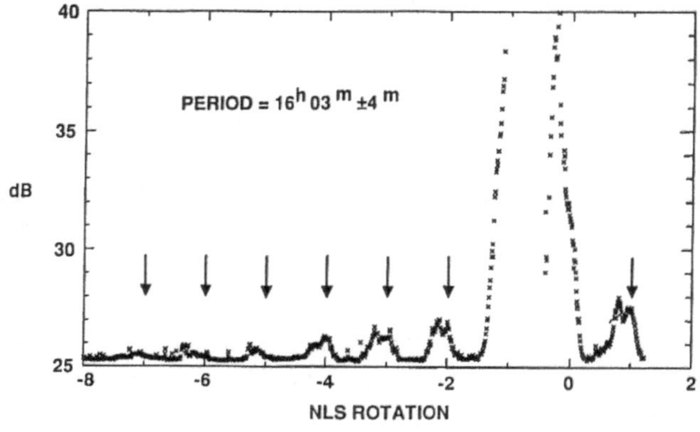

Fig. 4. Several cycles of radio bursts observed from Neptune plotted versus the Neptune rotation period indicated in the figure (Warwick, 1989). Subsequent analysis of a much larger body of data has established a more accurate period of 16.11 ± .05 hours.

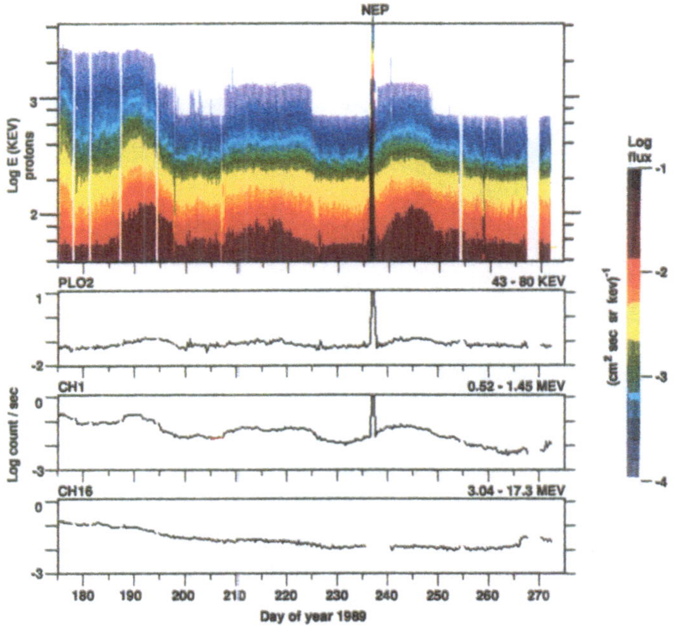

Fig. 3. The interplanetary radiation environment at Voyager 2 from late June through the end of September 1989 in 6-hour averages. The top panel shows a spectrogram with background subtracted to highlight intensity changes, while the bottom three panels represent count-rates from selected energy channels over a range in energy of a factor ~ 100. The intensity increase associated with the Neptune encounter is marked on top.

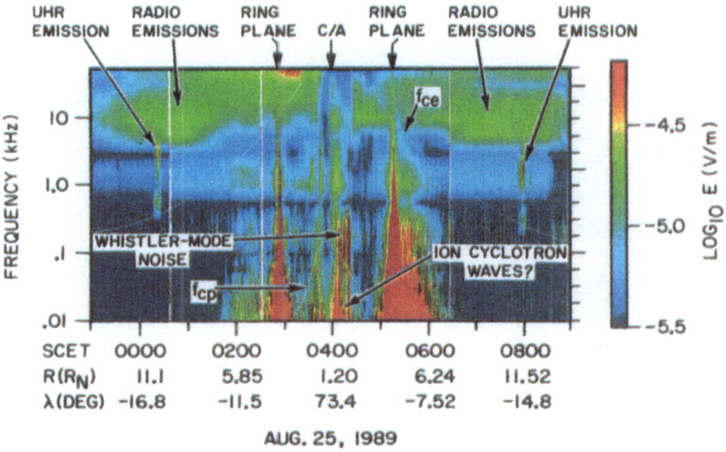

Fig. 9. The electric field intensities observed in the inner magnetosphere presented in a frequency-time spectrogram. The profiles of f_{ce} and f_{cp} are derived from the magnetic field data (Gurnett et al., 1989).

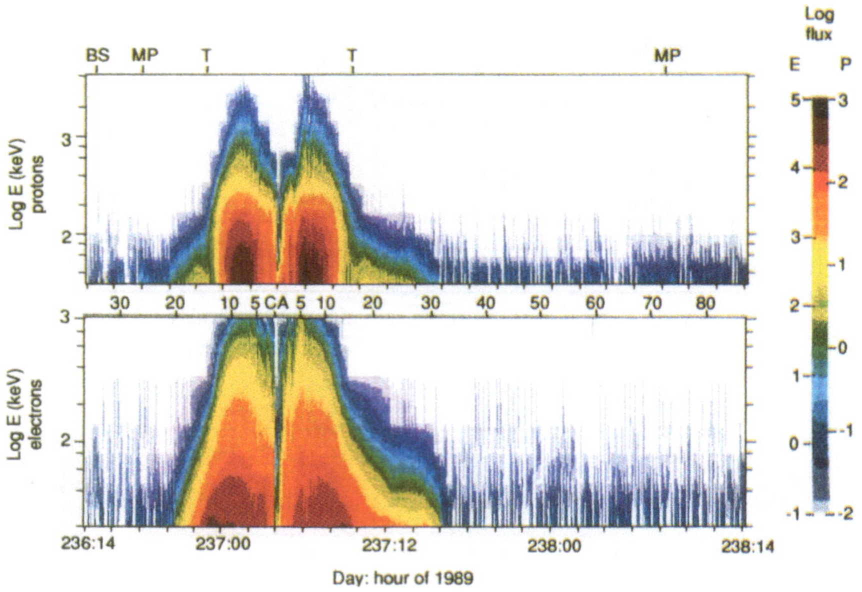

Fig. 10. Spectrogram of energetic ions and electrons over the indicated time interval. The bow shock (BS), magnetopause (MP), and Triton (T) orbit crossing times are shown on top, while radial distance in units of planetary radii are noted in the middle ($1R_N = 24{,}765$ km) (Krimigis et al., 1989).

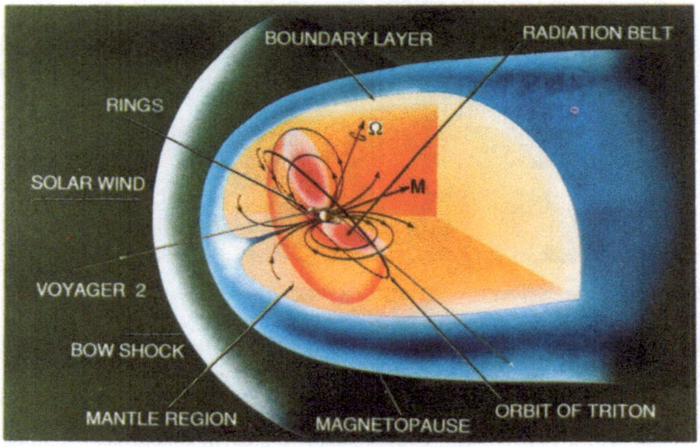

Fig. 19. Conceptual model of Neptune's magnetosphere as it might appear at the time of Voyager's entry through the cusp region. Principal features include the mantle region of semi-trapped particles, the radiation belt delineated by the orbit of Triton, and a boundary layer on the nightside (Krimigis et al., 1989).

encounter period of 17h 50m. Subsequent analysis has shown that the bursts were present as early as 30 days prior to closest approach and at least 22 days after closest approach (Warwick et al., 1989). The analysis of the more expanded body of data has resulted in a planetary rotation period of 16.11 ± 0.05h. In addition to the burst episodes, smooth emissions were detected in the frequency range from 20 kHz to 865 kHz for at least 10 days surrounding closest approach (Warwick et al., 1989). The overall spectrum of Neptune's radio emission is similar to that of Uranus, although the peak power is significantly lower than that of Uranus.

Magnetic Field. Crossing of the planetary bow shock occurred at 1438 spacecraft event time (SCET) on day 236 at a distance of ~ 34.9 R_N. The magnetic field signature is shown in Figure 5. The data are plotted versus Pacific Daylight Time (one-way light-time from the spacecraft was 4 hours 6 minutes at this time) and show a typical bow shock signature with significant turbulence inside the magnetosheath. Note that the interplanetary magnetic field upstream from the shock was generally along the spiral direction, with possible connection to the bow shock only after the spacecraft came close to the shock, as can be deduced from the geometry shown in Figure 2. Evidence for the presence of the shock was actually obtained about one hour 40 minutes earlier by the presence of electron plasma oscillations observed with the plasma wave experiment (Gurnett et al., 1989). The magnetopause crossing occurred at ~ 1800 SCET at a distance of ~ 26.5 R_N (Krimigis et al., 1989; Belcher et al., 1989; Ness et al., 1989), i.e. substantially later than expected from normal scaling of the magnetopause-bow shock distance. Analysis of the magnetic field data following closest approach soon disclosed the reasons why that was the case.

Figure 6 shows the results of the initial analysis of the magnetic field data. The Neptunian field is represented well by an offset tilted dipole (OTD) displaced from the center of Neptune by a large offset of ~ 0.55 R_N and inclined by 47° with respect to the rotation axis. The OTD moment is 0.133 Gauss-R_N^3 (Ness et al., 1989), and is a good representation of the field in the range of 4-15 R_N. Inside that range, the magnetic field is more complex, exhibiting two peaks near periapsis with a maximum observed field of ~ 10,000 nanotesla. The Neptune magnetic

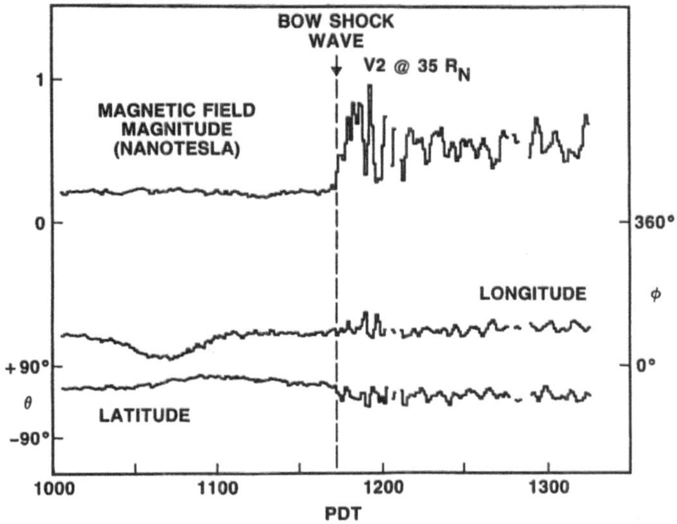

Fig. 5. Crossing of the Neptunian bow shock as observed by the magnetometer instrument plotted versus Pacific Daylight Time. The one-way light-time from the spacecraft at this time was four hours and six minutes. The field longitude is measured in a plane parallel to the sun's equatorial plane where 0° is anti-sunward and the latitude is positive northward (Ness, 1989).

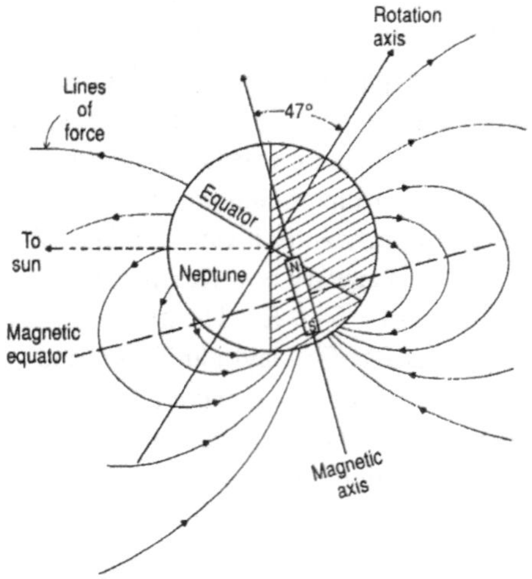

Fig. 6. Diagram of the OTD field lines of Neptune in the meridian plane containing the OTD center and the rotation axis. The OTD model has a moment of 0.133 Gauss-RN3, a dipole tilt of 47° and offset from the planet's center of 0.55 RN (0.17, 0.46, and -0.24 RN in a right-handed coordinate system in which the positive Z-axis is aligned with the rotation axis and the X-axis passes through the 0 meridian) (Ness et al., 1989).

field orientation resulted in a rather complex trajectory by the Voyager spacecraft in terms of magnetic coordinates.

In figure 7 the coordinates calculated from the OTD model are used to show the spacecraft trajectory in L (upper panel) and magnetic latitude (lower panel). The dashed line at lower L values represents radial distances inside ~ 4 R_N where the OTD model is not expected to be valid. The figure shows that the spacecraft crossed the magnetopause in a latitude range (≤ - 50°) which may well correspond to the polar cusp region of the Neptunian magnetosphere. This would be the first such occurrence in a planetary encounter by the Voyager spacecraft and may explain the extended and rather unusual magnetosheath region, and the relative delay in encountering the magnetopause. The figure also shows that the spacecraft crossed the magnetic equator at least three times inside 20 R_N and at least once outside that distance. Particularly noted is the time of 0420 SCET when the spacecraft may have crossed the magnetic equator after closest approach; it will be referred to in the data presentation in the next several sections. The subsequent discussion of the plasma, plasma wave, and energetic particle data makes extensive reference to the trajectory of the spacecraft as depicted in Figure 7.

Plasma and Plasma Waves. The Voyager plasma instrumentation measures plasma ions and electrons over the range ~ 10 eV to ~ 6 keV. A summary of this data is shown in Figure 8. The top two panels show the current due to ions and electrons, respectively, for four days surrounding closest approach in units of femptoamperes (10^{-15} amperes), while the bottom panel shows the electron density (Belcher et al., 1989). Note that the instrument background is in the range of 200-300 fA and that the currents inside the magnetosphere were quite low, with the exception of the plasma enhancements (PE); the corresponding electron densities are shown in the bottom panel. Immediately upon crossing the bow shock solar wind electrons were sufficiently heated to bring them above the low threshold of the plasma detector at 10 eV, but the intensities dropped following the magnetopause crossing because of the low density of magnetospheric electrons. Maximum densities were ~ 0.1 within the magnetosphere, although at some of the equatorial crossings the ion densities were ~ 1 cm^{-3} (Belcher et al., 1989). Transition into the outbound magnetosheath is clearly evident in both the ions and electrons, as is the transition to the interplanetary medium following the bow shock crossing at ~ 2000 SCET on August 27.

The principal findings from the plasma wave detector are summarized in spectrogram form in Figure 9 over the frequency range from 10 Hz to 50 kHz for ~ 10 hours surrounding closest approach (Gurnett et al., 1989). The most intense emissions occurred at the lower

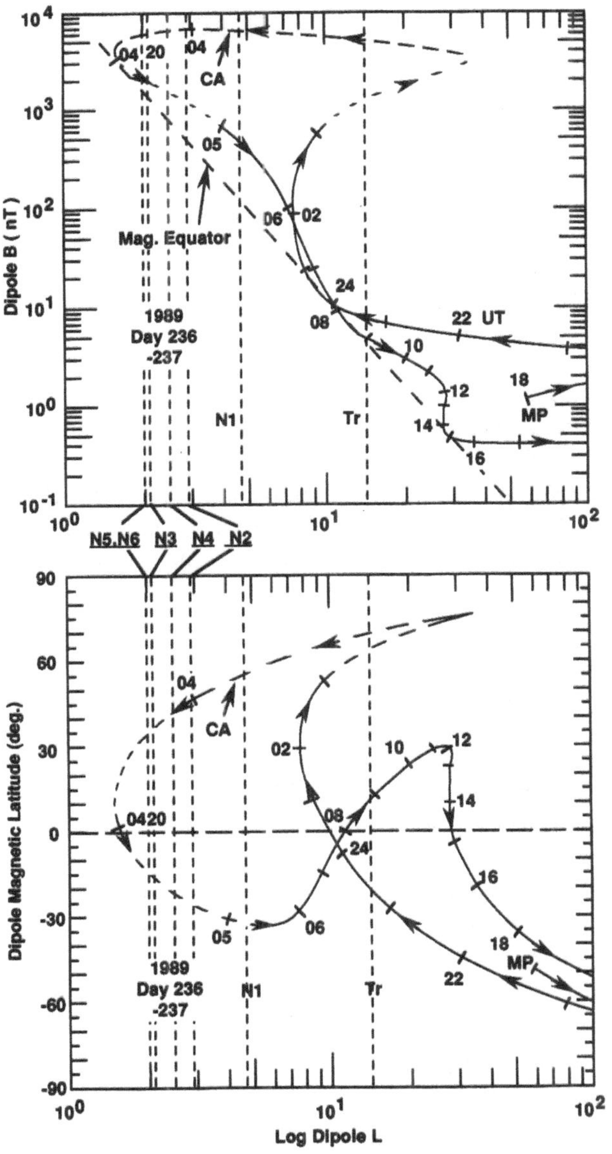

Fig. 7. Trajectory of the Voyager-2 spacecraft at Neptune in magnetic coordinates, based on the OTD model. The range over which the OTD model is not expected to be valid is shown by the dashed curve. Vertical lines show the minimum L-shells of Triton and of the five newly-discovered Neptunian moons.

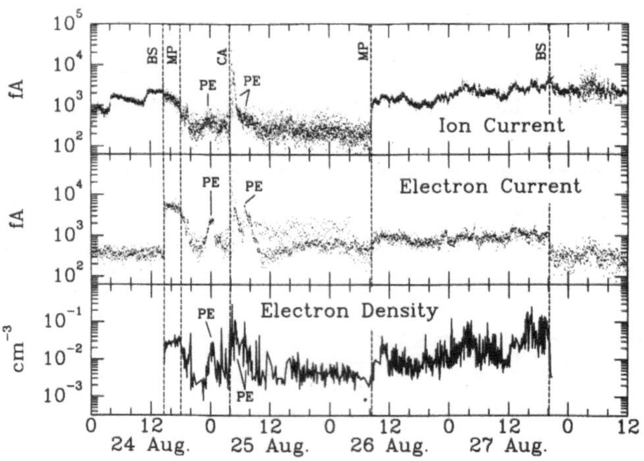

Fig. 8. Measurements of ion and electron current (energy range, 10 eV to 1 keV for ions and 10 eV to 140 eV for electrons) and electron density (bottom panel) observed by the PLS experiment near Neptune. The current is in femptoamperes (10^{-15}A). The various boundary crossings are indicated as is the plasma enhancements (PE) generally associated with the magnetic equator (Belcher et al., 1989).

frequencies at the time of ring plane crossings, both inbound and outbound, and are due to the bombardment of the spacecraft with micron-size particles through the well-known mechanism of vaporization on the body of the spacecraft, ionization, and collection of charge on the experiment antennas (Scarf et al., 1982). The corresponding impact rates were ~ 250 impacts per second at the inbound ring crossing and ~ 100 per second at the outbound crossing. In addition to the radio waves evident at the higher frequencies, there were distinct upper hybrid resonance (UHR) emissions at the times of the inbound (0025 SCET) and outbound (0800 SCET) crossings of the magnetic equator. Since such waves occur at the upper hybrid resonance frequency given by $f_{UHR} = (f_{ce}^2 + f_{cp}^2)^{1/2}$ where the electron cyclotron frequency f_{ce} can be computed from the measured magnetic field strength while the electron plasma frequency f_{cp} is related to the density, it is possible to compute the electron density in these regions. The resulting numbers are in the range of 0.12 cm^{-3} and 0.04 cm^{-3} at the inbound and outbound crossings, respectively, i.e. similar to the electron densities measured by the plasma instrument.

There is a range of complex plasma wave emissions near closest approach which are difficult to interpret at the present time. Most of these occur immediately after closest approach (~ 0356 SCET) and terminate at ~ 0430 SCET. Note that emissions from 10-31 Hz are below the proton-cyclotron frequency and could be due to electrostatic or electromagnetic ion cyclotron waves (Gurnett et al., 1989). Despite the appearance of apparent whistler mode noise, the authors do not feel that they have observed whistlers originating from lightning during this encounter (Gurnett et al., 1989).

Energetic Particles. The principal observations of energetic particles are summarized in the spectrogram shown in Figure 10, for ions (top panel) and electrons (bottom panel) (Krimigis et al., 1989). Particle signatures at the marked bow shock (BS) and, most importantly, magnetopause crossings (MP) are evident. The intensities, however, did not increase sharply upon crossing the magnetopause, as has been the case in previous planetary encounters; instead they remained low until ~ 2000 SCET on day 236 when a gradual rise began as the spacecraft resumed its approach to lower magnetic latitudes (Figure 7). Note that the next intensity minimum occurs at ~ 0400 SCET, near closest approach, while a third relative minimum occurs at ~ 1200 SCET on day 237, i.e. the three decreases appear to be separated by ~ half the planetary rotation period of 8 hours.

The principal increases in intensities ≥ 200 keV occurred inside the orbital radius of Triton, marked on top of the figure. In addition, there appears to be a relative minimum in the proton intensity near the orbit of Triton, inbound and outbound, at the lower energies. The overall intensity profile is basically symmetric with respect to closest approach, with the possible exception of an extended lower energy region outbound after ~ 1200 SCET, potentially associated with a boundary layer as the spacecraft exited into the southern tail lobe.

Electron intensities extended to energies ≥ 1 MeV as shown in Figure 11 (Stone et al., 1989). The top panel shows that intensities of energetic electrons extend at least ≥ 2.5 MeV, but not higher than 5 MeV, and exhibited considerable structure around the time of closest approach to the planet. The bottom panel shows the spacecraft trajectory in L, and marks the crossings of satellites 1989N1, N2, and of Triton. Curves C, B and F refer to alternative values of L that may account plausibly for the observed features denoted in the top panel in a region where the OTD model (curve D) may not be representative of the real magnetic coordinates. An extended discussion of these features in Figure 11 can be found in the reference (Stone et al., 1989).

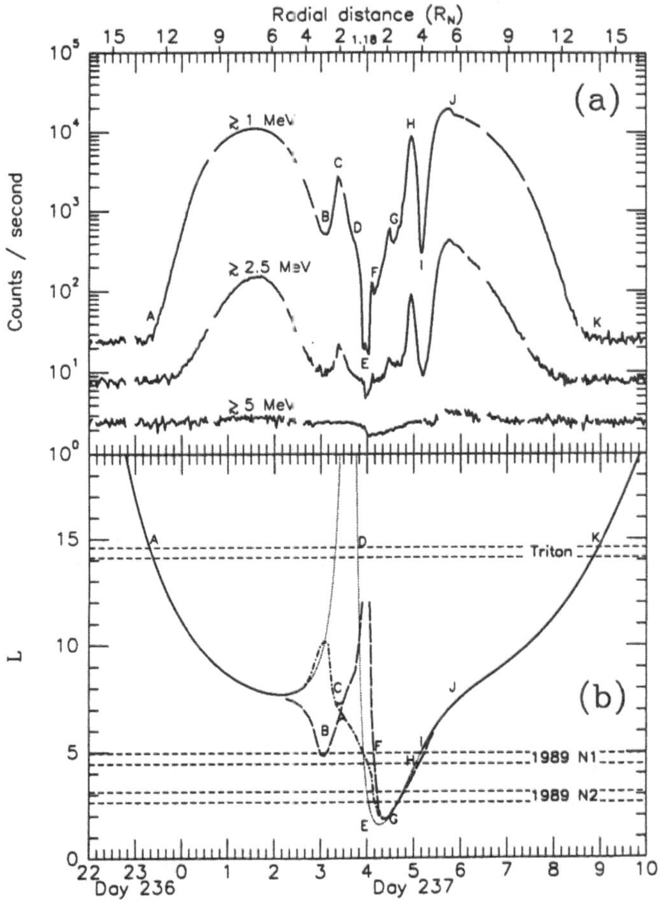

Fig. 11. Count rate profiles for electrons with indicated kinetic energies versus spacecraft event time (SCET). The bottom panel shows the position of the spacecraft versus L based on the OTD model, together with possible alternatives at close-in distances depicted by curves B, C and F (Stone et al., 1989).

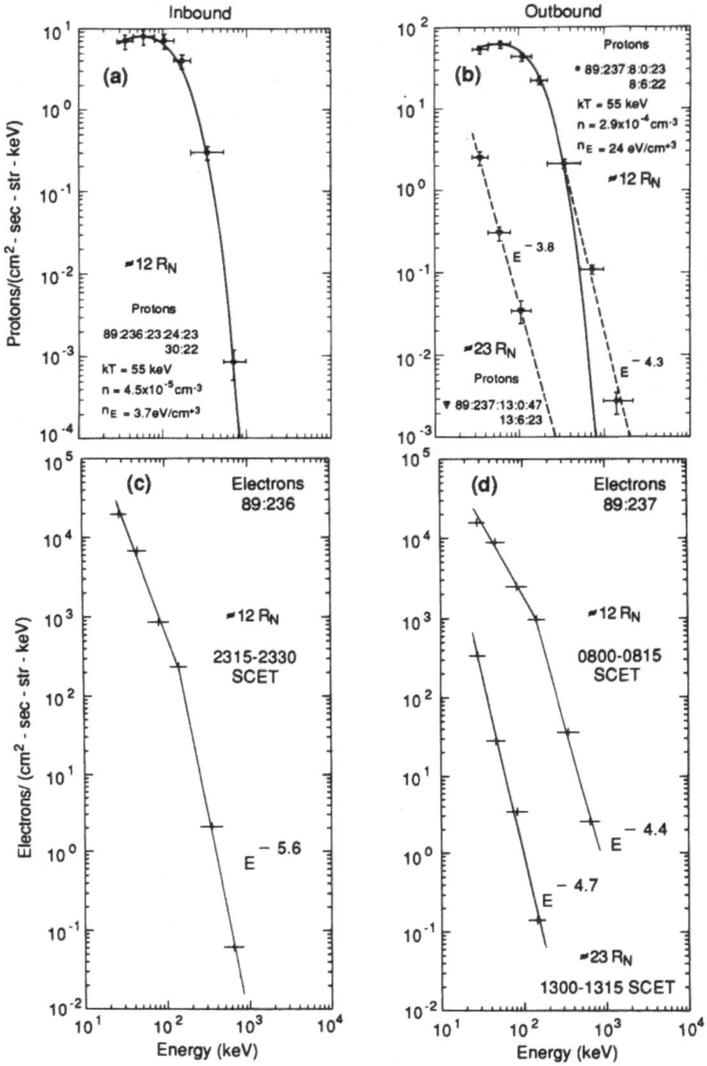

Fig. 12. Differential energy spectra of ions (top panels) and electrons (bottom panels) at selected times inbound and outbound. Note the difference in functional form inside and outside Triton's orbit (right hand panels) (Krimigis et al., 1989).

Examples of some characteristic differential energy spectra for protons and electrons are shown in Figure 12 (Krimigis et al., 1989). The inbound proton spectrum (panel a) inside the radial distance of Triton is well represented by a Maxwellian distribution over the energy range from ~ 30 keV to ~ 1 MeV. The distribution is characterized by an exceptionally hot temperature (kT ~ 55 keV) and very low number density. Panel (b) shows spectra from the outbound trajectory, including the proton spectrum outside the Triton radial distance (~ 1300 SCET); it is a simple power with $\gamma \sim 4$. Obviously there is a drastic difference in spectral form between outside and inside the Triton L-shell, indicating that Triton plays a determining role in controlling the outer regions of the Neptunian magnetosphere. Panels (c) and (d) display spectra for the electrons, inbound and outbound, respectively. These spectra can be described by simple power laws, being steeper at energies ≥ 200 keV, except inside the orbit of Triton where a single power law describes the spectrum with an effective cut-off at ~ 200 keV (d).

Figure 13 shows corresponding energy spectra for protons (left panel) and electrons (right panel) at energies ≥ 1 MeV. Evidently the spectra for both protons and electrons continue to fall steeply above this energy with an effective cut-off at about 5 MeV in both cases. Note that at a given energy the fluxes of protons are substantially lower than those of the electrons throughout the magnetosphere.

Composition. Composition measurements were performed by Voyager instruments in Neptune's magnetosphere in both energetic particles (LECP) and plasma (PLS) energies. Figure 14 shows a histogram of pulse height measurements from LECP when Voyager was within 11 R_N of the planet. A proton peak is clearly evident, showing protons to be the dominant ions present. In addition, there is a shoulder in the distribution at the expected location of H_2 molecular ions. Further, there is a small peak containing three counts at the location of ^4He. The measured abundances at equal energy per nucleon intervals (0.57 to 1.0 MeV/nuc) are $H:H_2:{}^4He$ in the proportion 1300:1:0.1 (Krimigis et al., 1989). Such high proton-to-helium ratios rules out the solar wind (where $H:{}^4He$ ratio is typically ~ 15) as an important source of the plasma in Neptune's magnetosphere.

Figure 15 shows data from the PLS instrument versus energy/charge taken at the time of the first equatorial crossing on day 237. The data are best fit by 2 isotropic Maxwellians, assuming corotation of the plasma, that represent H^+ and N^+ ions. The corresponding densities and temperatures are 0.7 cm^{-3} and 7 eV for the protons, and 0.04 cm^{-3} and 65 eV for N^+ (Belcher et al., 1989). The obvious source for the N^+ component is Triton's upper atmosphere which consists mostly of nitrogen as shown by the UV instrument (Broadfoot et al., 1989). Overall

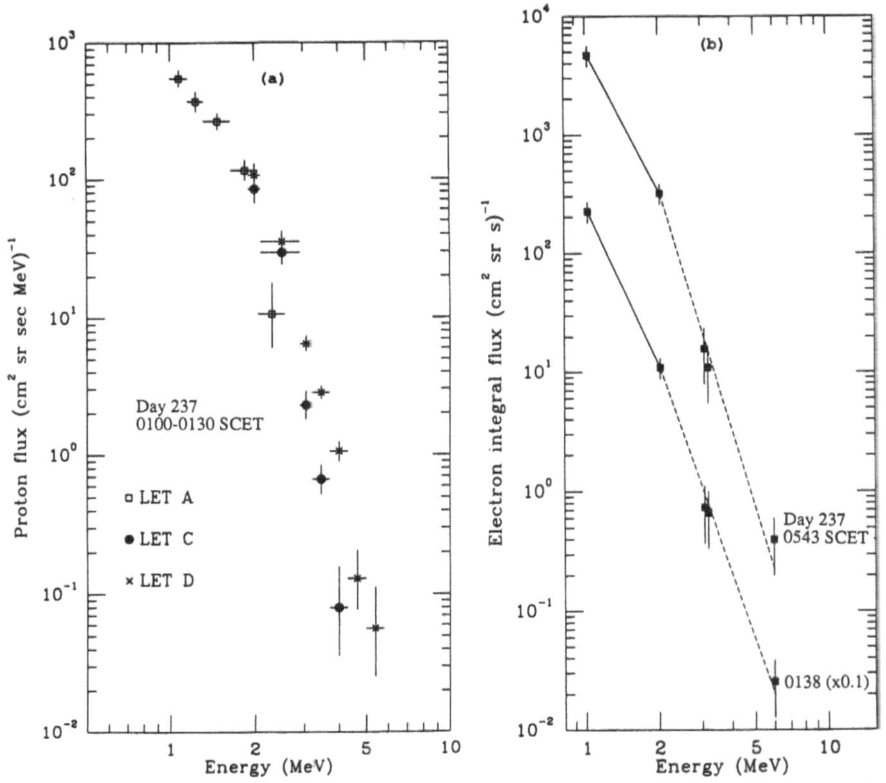

Fig. 13 (a). Proton differential intensity spectra in three telescopes during the period 0100 to 0130 SCET on day 237. (b) Selected electron energy spectra for 15-minute intervals centered at the indicated times. These energies are higher than those shown in Figure 12 (Stone et al., 1989).

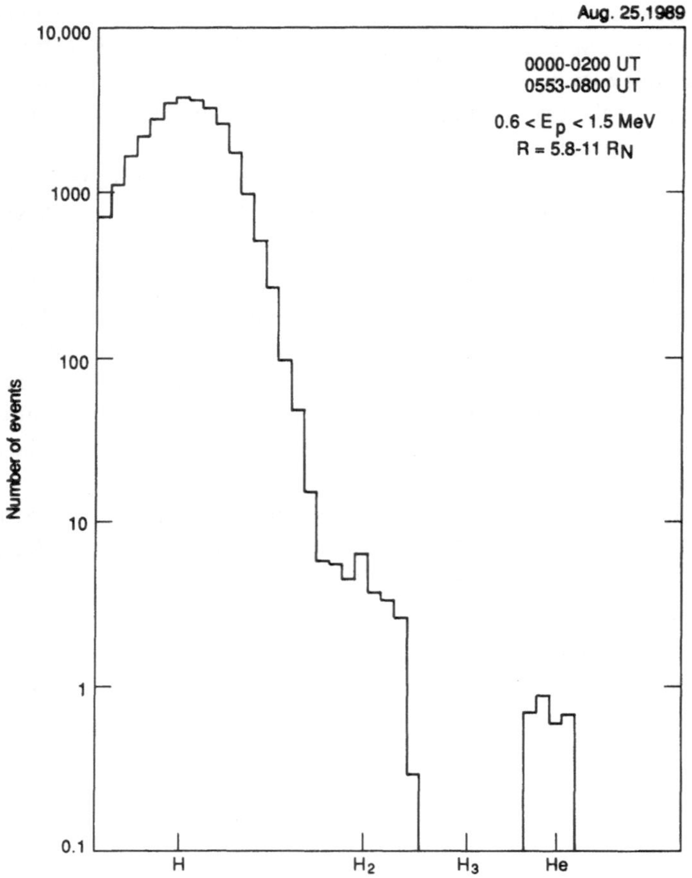

Fig. 14. Logarithmic histogram of pulse-height analyzed events in the Low Energy Particle Detector telescope system of LECP. The shoulder to the right of the proton distribution indicates the presence of H_2^+ ions in the magnetosphere of Neptune (Krimigis et al., 1989).

the composition measurements of the Voyager 2 instruments strongly suggest that the plasma and energetic particle population in the magnetosphere of Neptune is indigenous to the planet and that the solar wind component is of minor importance at best.

Aurora on Neptune? A principal feature of the Voyager 2 encounter with Neptune was the close flyby distance of ~ 4900 km over the planet's north pole. This flyby geometry offered the opportunity potentially to examine the high latitude extent of any radiation belts and to measure precipitating electrons and protons which might produce possible auroral emissions. Figure 16 shows count-rate profiles (bottom panel) of selected electron and ion channels together with the spectral index for low energy electrons (top panel) during the time interval surrounding the polar passage of the spacecraft (Krimigis et al., 1989). Abrupt discontinuities in the curves, for example at 0435 and 0441 SCET are due to the two different "parking" positions of the LECP aperture that are 90 degrees apart in the LECP's scan plane (for more details, see Krimigis et al., 1989).

In the lowest, proton, curve, there are two sharp intensity changes at ~ 0325 and ~ 0510 SCET; these are reminiscent of the latitudinal extent of trapped protons in the Earth's outer radiation belt. The second intensity curve from the bottom exhibits a behavior similar to that expected from outer zone radiation belt electrons at Earth, i.e. a gradual increase occurs near the location of the high energy proton dropouts. Such an interpretation is consistent with the trajectory in B, L space depicted in Figure 7. Also marked in the figure are the tentative identifications of satellite 1989N1 with a clear-cut signature seen at 0510 SCET and a potential signature also observed at ~ 0308 SCET. Reference to Figure 7 suggests that the spacecraft may not have crossed the minimum L shell of 1989N1 at 0308 SCET, but the signature in the particle data indicates a potential crossing. The particle signatures can be used in this manner to augment the magnetic field data in arriving at an appropriate model for the magnetic field configuration in the inner magnetosphere (e.g. Van Allen et al., 1980).

There is considerable and intermittent activity in the low energy electron and proton intensities (second and third curves from the top, respectively) with as many as four distinct spikes in the low energy electron profile and one proton spike between ~ 0350 and ~ 0405 SCET. In addition, there is a spike in the proton intensity at ~ 0420 SCET, which is followed by an apparently similar spike of low energy electrons approximately two minutes later. This proton pulse coincides with a clear deficit in the intensity of ~ 250 keV electrons. We note that this occurs at the time when, according to the OTD field model, the spacecraft was expected to cross the magnetic equator (Figure 7). A distinctive feature of these electron spikes is readily evident in the spectral variations shown in the top panel. The electron spectrum became quite soft for the period of 0350 to 0425 SCET, not unlike spectral variations which are observed in the auroral zones of Earth.

The overall impression from the high latitude trajectory of Voyager 2 is that of the passage of a spacecraft from a region of magnetically trapped radiation into a magnetic polar cap region. Such a passage at Earth also occurs with considerable electron and ion precipitation activity and a softness of the electron spectrum on either side of the proton trapping boundary. It is also possible that the spacecraft traversed the region which may map into the location of Triton's flux tube in the outer magnetosphere of Neptune. The spacecraft also traversed a high magnetic field region during its passage (Figure 7), so that any precipitation pattern may be complicated by the equivalent of a "South Atlantic anomaly" region in the Earth's radiation belts.

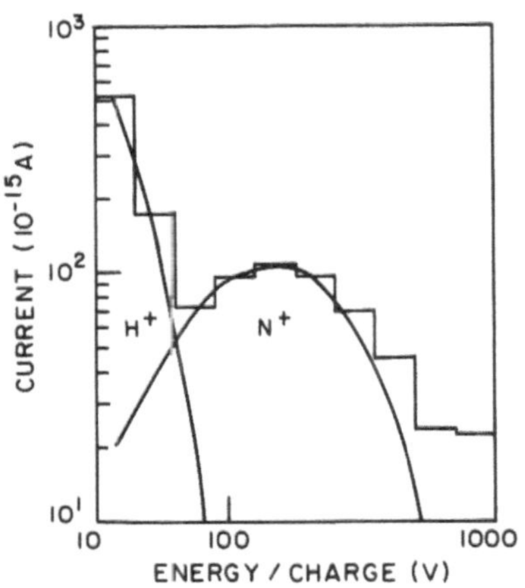

Fig. 15. Ion spectrum from the first plasma sheet crossing taken at 001800 SCET on day 237, with current into the D-sensor plotted versus energies. The curves show the best fit to the data assuming two ion species, protons and N^+, and a convected isotropic Maxwellian velocity distribution at corotation velocity (Belcher et al., 1989).

Auroral emissions from the dark side of Neptune principally in the wavelength of H Lyman β (1025 Å) were detected by the UVS experiment (Broadfoot et al., 1989). The longitudinal dependence of the emission is shown in Fig. 17 and exhibits peaks near ~ 30° and ~ 200°. The radiated power in UV from one hemisphere is ~ 2.5 x 10^7 W. The estimated energy input rate from the electrons and protons observed by LECP is ~ 3 x 10^7 W. It is expected that conversion efficiency from particle to UV excitation energies is at most ~ 0.1, so that one would expect the particle energy input to exceed 3 x 10^8 W. It is important to note, however, that the UV measurements are global in nature, whereas the particle measurements were made in a limited part of the polar cap region. If, for example, the flux tube threading through Triton represents the principal particle input into the aurora of Neptune, then the electron intensities

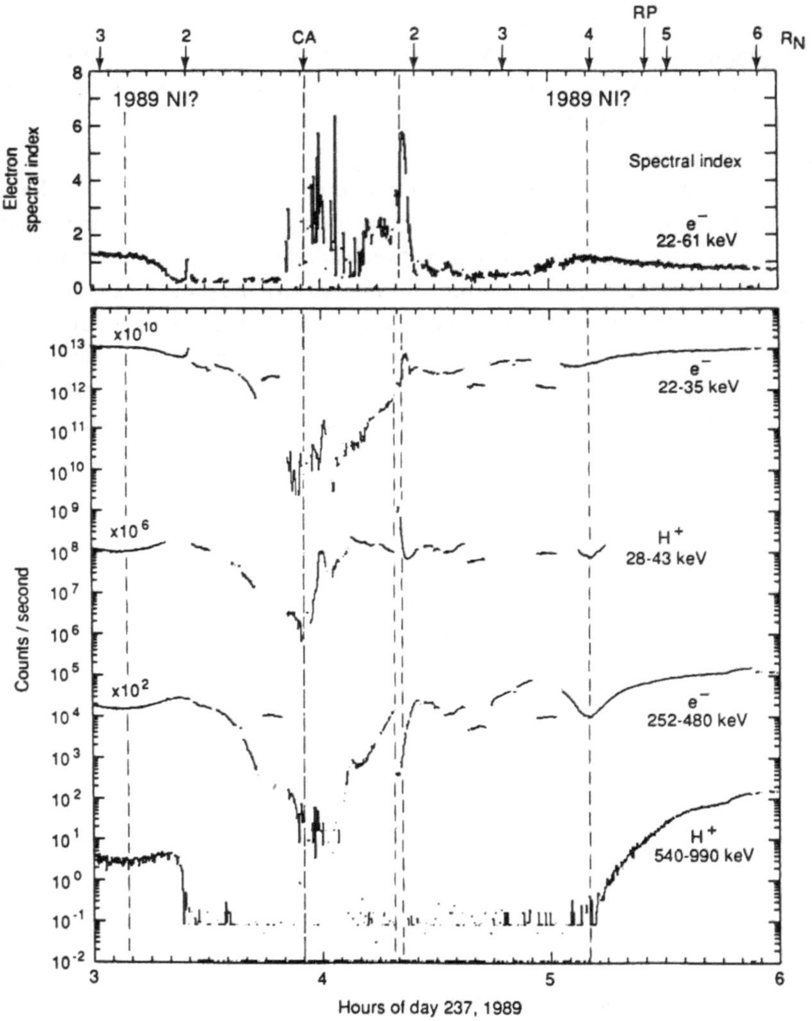

Fig. 16. Counting rates of selected channels at 12 sec. time resolution from the LECP instrument. A radial distance scale in R_N and the ring plane (RP) crossing outbound are noted on top. An intensity signature most likely due to satellite 1989N1 is marked. Vertical lines at closest approach and at the time of the ion-electron spike at ~ 0420 SCET are drawn to guide the eye. The top panel is a fit to a power law spectrum at the two lowest energy channels (Krimigis et al., 1989).

are larger by at least a factor of 100 and would represent an energy input in excess of 10^9 W, i.e. more than sufficient to account for the global auroral emissions observed by the UV instrument.

In fact, Broadfoot et al. (1989) have suggested that this may be the case. Their qualitative model is shown in Figure 18. They suggest that each auroral region is formed by a partial torus due to gas escaping from Triton's atmosphere and becoming ionized principally at the crossings of the magnetic equator which occurs twice per planetary rotation. The plasma will populate the region stretching along the magnetic equator and would produce aurora on Neptune over the longitude intervals observed, as shown in Figure 18. As indicated above, the low particle fluxes observed by the LECP instrument could be due to the fact that the Voyager spacecraft only made a partial traversal of an auroral arc, or that the altitude was low enough such that significant precipitation might have occurred above the spacecraft due to the magnetic anomaly in the planet's field.

3. DISCUSSION

The principal findings from the investigation of the magnetosphere of Neptune by Voyager 2 are summarized in Figure 19 (Krimigis et al., 1989). It shows the higher energy part of the radiation belt principally confined within the orbit of Triton. The orientation of the magnetic moment is drawn at its likely position at the time of the magnetopause crossing by the Voyager 2 spacecraft. The 47° dipole tilt angle with respect to the rotation axis of the planet enabled the spacecraft to cross the magnetopause inbound while the magnetosphere was in a "pole-on" configuration. Such was the configuration anticipated for Uranus prior to Voyager encounter (Siscoe et al., 1975), but not actually achieved because of the large (60°) Uranian dipole tilt and the peculiar orientation of the Uranian rotation axis pointing nearly toward the Sun. The nearly pole-on configuration at Neptune occurs once per Neptune rotation, so it was fortuitous that Voyager at Neptune made the first measurements in the cusp region of an outer planet magnetopause. The large Neptunian dipole tilt also implies that the satellites of Neptune, like those of Uranus, trace complicated trajectories in the magnetic field, each satellite encountering a large range of L shells and magnetic latitudes (Paonessa and Cheng, 1987).

Preliminary calculations of plasma stresses in the magnetospheres of Neptune from the LECP data (Krimigis et al., 1989) show that maximum pressure along the Voyager trajectory was found near the three close-in equatorial crossings (i.e. at ~ 0030, ~ 0800, and ~ 1600 SCET, all on day 237). The ratio β of plasma to magnetic pressure at Neptune was ~ 0.2 at all three equatorial crossings, similar to that observed at the one encounter of the magnetic equator of Uranus of $\beta \sim 0.13$. These low values of β imply that at Neptune, like at Uranus, the local plasma stresses do not significantly distort the magnetic field from the vacuum configuration. This contrasts with the situation at Jupiter and Saturn where $\beta > 1$ for most of the plasma sheet encounters, and a strong magnetodisk distortion of the magnetic field results.

Spectral and anisotropy signatures in the magnetosphere of Neptune are similar to those of Saturn with the moon Triton playing a role similar to that of Titan. For example, the energy spectra outside the orbit of Triton are quite soft with the high energy proton and electron intensity gradients occurring just inside the orbits of Triton, both inbound and outbound. Typically energetic particles are lost due to wave-particle interactions in cold plasma density gradients (Cheng et al., 1985), for example the case of Jupiter's Io plasma torus (Thomsen et al., 1977). A heavy ion plasma, presumably from Triton has been detected as shown in Figure 15 (Belcher et al., 1989), but Triton's interaction with the magnetosphere is not fully understood as yet.

The interaction of the trapped particles observed by the LECP with a hypothetical Triton torus could drive the auroral processes that apparently occur over Neptune's high latitude region. In fact, the particle observations shown in Figure 14 are consistent with a power input of ~ 5 x 10^{-4} ergs/cm^2 sec. The total power input into the atmosphere of Neptune is of order ~ 3 x 10^7 W if one assumes that the aurora zone has a radius of ~ 10°. Such a power input is surprisingly small compared to those at Jupiter, Saturn and Uranus but, as indicated earlier, it is possible that a combination of altitude and latitude effects owing to the anomalous magnetic field of Neptune could have severely altered the fluxes reaching the Voyager spacecraft. The probability that the observed soft electron and ion fluxes over the polar cap region are evidence of auroral emissions is also supported by the fact that the plasma wave experiment observed hiss and chorus emissions associated with that same region during closest approach (Gurnett et al., 1989).

Several additional results have emerged from this preliminary analysis of particles and fields data during the encounter with Neptune's magnetosphere. For example, the putative observation of synchrotron emission from Neptune (de Pater and Goertz, 1989) is unlikely to be verified, as indicated by the very low fluxes of relativistic electrons observed near closest approach during the passage of Voyager 2. Another result is the failure to detect either upstream ion events or energetic charge exchange neutrals outside the Neptunian magnetosphere. This contrasts with the situation at Earth, Jupiter and Saturn where both upstream ion bursts and energetic charge exchange neutrals have been detected at great distances (Krimigis, 1986). At

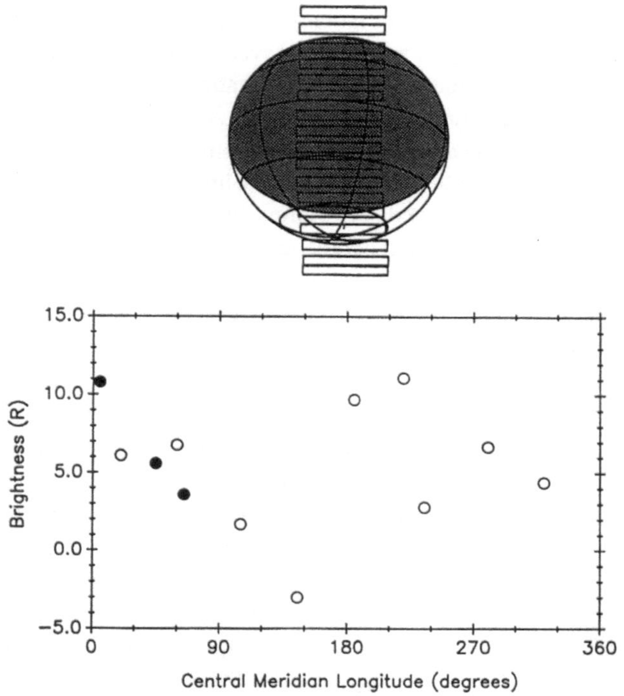

Fig. 17. Neptune emission versus longitude of the central meridian from the dark side as seen by the UVS instrument. The emission has been integrated in the range 967 ≤ λ ≤ 1115 Å from twelve successive observations of the non-illuminated portion of Neptune. The filled points are from the beginning of a second rotation which reproduces the enhancement peak at ~ 30° (Broadfoot et al., 1989).

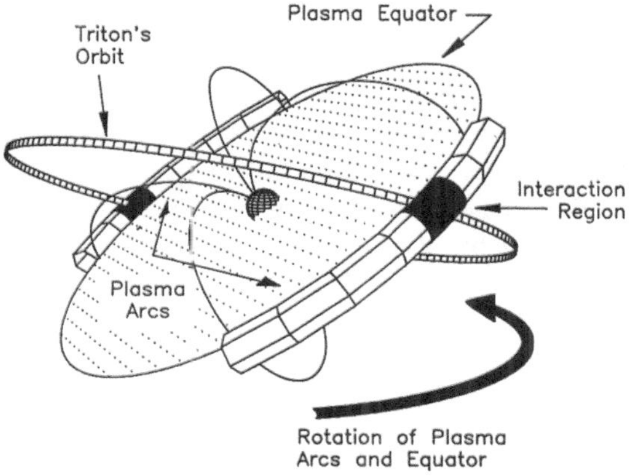

Fig. 18. Model of the injection of gas from the atmosphere of Triton that populates segments of the magnetic equatorial plane after ionization and feeds Neptune's aurora (Broadfoot et al., 1989).

Uranus, on the other hand, upstream ion events but not charge exchange neutrals were detected (Krimigis et al., 1988). Since upstream ion events are observed only when the interplanetary magnetic field connects the spacecraft to the bow shock, and since prior to the Neptune encounter the IMF lay near the nominal spiral direction (Figure 2) it is likely that this is the reason for the absence of such events. Furthermore the escape of magnetospheric particles, the dominant process responsible for upstream events, occurs primarily at low magnetic latitudes; Voyager apparently entered and left the Neptune magnetosphere at very high latitudes. The absence of energetic neutrals on the other hand, implies a small or non-existent atomic H exosphere in the vicinity of the planet. Having measured the intensity of energetic ions at Neptune it is possible to place an upper limit on the volume-averaged atomic H density in this region of $\leq 20\ cm^{-3}$ within a volume $< 6\ R_N$ in radius. Such an upper limit is consistent with the absence of an extended atomic H corona on Neptune reported by the UVS instrument (Broadfoot et al., 1989).

Finally, charged particle bombardment of satellites and rings can produce a variety of chemical and physical alterations of solid surfaces (Lanzerotti et al., 1988; Cheng et al., 1986). Both sputtering and discoloration can occur over time, depending on the composition of the surface. Preliminary estimates on the rate of darkening for a surface containing methane ice at the location of 1989N1 around Neptune have been made (Krimigis et al., 1989). The results show that darkening to an albedo of perhaps 10% will occur to a depth of ~ 1 micron within < 10^5 years. Similar darkening to 10 microns will occur within < 10^6 years. Thus, the dark materials of the rings and ring particles and inner satellites of Neptune (Smith et al., 1989) could have been introduced within the last million years through radiation processing by the Neptune magnetosphere just as easily as in the earliest days of the solar system. It should be pointed out that radiation effects can also become important in interactions between electrons and protons with the atmosphere of Triton. These effects, however, have yet to be quantified in detail.

The Voyager encounter with Neptune brings to a close a chapter of exploration of planetary magnetospheres for the remainder of the twentieth century. It is only fitting that this coincides with our gathering here in Stockholm to honor the person who pioneered this research, Professor James A. Van Allen. His leadership, his seminal contributions to the field, and his contributions to educating a new generation of researchers will live as a shining example for the rest of us to emulate as we pursue our professional careers in the years to come.

ACKNOWLEDGEMENTS

This work draws heavily from the work of the LECP science team as reported in the journal Science on December 15, 1989. I am indebted to my colleagues for many discussions regarding the contents of this review. I am also grateful to my colleagues on the Voyager Science Team, Drs. J. W. Belcher, D. A. Gurnett, N. F. Ness, A. L. Broadfoot and J. Warwick for making their results available prior to publication and for many useful discussions. This work has been supported by NASA under Task I of Contracts N00039-89-C-5301 and N00039-89-C-0001.

REFERENCES

Belcher, J.W., H.S. Bridge, F. Bagenal, B. Coppi, O. Divers, A. Eviatar, G.S. Gordon, Jr., A.J. Lazarus, R.L. McNutt, Jr., et al., 1989, Plasma observations near Neptune: Initial results from Voyager 2, *Science,* 246, 1478.

Broadfoot, A.L., S.K. Atreya, J.L. Bertaux, J.E. Blamont, A.J. Dessler, T.M. Donahue, W.T. Forrester, D.T. Hall et al., 1989, Ultraviolet spectrometer observations of Neptune and Triton, *Science,* 246, 1459.

Cheng, A.F., S.M. Krimigis, and T.P. Armstrong, 1985, Near equality of ion phase space densities at Earth, Jupiter, and Saturn, *J. Geophys. Res.*,90, 526-530.

dePater, I., and C.K. Goertz, 1989, Synchrotron radiation from Neptune: Neptune's magnetic field and electron population, *Geophys. Res. Lett.,* 16, 97.

Gurnett, D.A., W.S. Kurth, R.L. Poynter, L.J. Granroth, I.H. Cairns, W.M. Macek, S.L. Moses, F.V. Coroniti, C.F. Kennel, D.D. Barbosa, 1989, First plasma wave observations at Neptune, 1989, *Science,* 246, 1494.

Krimigis, S.M., T.P. Armstrong, W.I. Axford, C.O. Bostrom, C.Y. Fan, G. Gloeckler, L.J. Lanzerotti, E.P. Keath, R.D. Zwickl, J.F. Carbary and D.C. Hamilton, 1979, Low-energy charged particle environment at Jupiter - A first look, *Science,* 206, 977.

Krimigis, S.M., 1986, Energetic ions upstream of planetary bow shocks: Fermi acceleration or leakage? *Comparative Study of Magnetospheric Systems,* CNES Editor, CEPADUES Editions, Toulouse, France, 99-124.

Krimigis, S.M., E.P. Keath, B.H. Mauk, A.F. Cheng, L.J. Lanzerotti, R.P. Lepping, and N.F. Ness, 1988, Observations of energetic ion enhancements and fast neutrals upstream and downstream of Uranus' bow shock by the Voyager 2 spacecraft, *Planetary and Space Sci.,* 36, 311-328.

Krimigis, S.M., T.P. Armstrong, W.I. Axford, C.O. Bostrom, A. F.Cheng, G. Gloeckler, D.C. Hamilton, E.P. Keath, L.J. Lanzerotti, B.J. Mauk, and J.A. Van Allen, 1989, Hot plasma and energetic particles in Neptune's magnetosphere, *Science,* 246, 1483.

Ness, N.F., M.H. Acuña, L.F. Burlaga, J.E.P. Connerney, R.P. Lepping, F.M. Neubauer, 1989, Magnetic fields at Neptune, *Science,* 246, 1473.

Ness, N.F., 1989, Press Conference, JPL, August 25, 1989.

Paonessa, M.T. and A.F. Cheng, 1987, Satellite sweeping in offset, tilted dipole fields, *J. Geophys. Res.*, 92, 1160-1166,

Scarf, F.L., D.A. Gurnett, W.S. Kurth, and R.L. Poynter, 1982, Voyager 2 plasma wave observations at Saturn, *Science,* 215, 587,

Siscoe, G.L., 1975, Particle and field environment of Uranus, *Icarus,* 24, 311.

Smith, B.A., L.A. Soderblom, D. Banfield, C. Barnet, A.T. Basilevksy, R.F. Beebe, K. Bollinger, J.M. Boyce, A. Brahic, G.A. Briggs et al., 1989, Voyager 2 at Neptune: Imaging science results, *Science,* 246, 1422.

Stone, E.C. and E.D. Miner, 1989, The Voyager 2 encounter with the Neptunian system, *Science,* 246, 1417.

Thomsen, M.F., C.K. Goertz, and J.A. Van Allen, 1977, On determining magnetospheric diffusion coefficients from the observed effects of Jupiter's satellite Io, *J. Geophys. Res.*, 92, 5541.

Van Allen, J.A., M.F. Thomsen, B.A. Randall, R.L. Rairden, C.L. Crosskreutz, 1980, Saturn's magnetosphere, rings and inner satellites, *Science,* 207, 415.

Warwick, J.W., D.R. Evans, G.R. Pelzer, R.G. Peltzer, J.H. Romig, C.B. Sawyer, A. C. Riddle, A. E. Schweitzer, M. D. Desch, M. L. Kaiser, W. M. Farrell et al., 1989, Voyager planetary radio astronomy at Neptune, *Science,* 246, 1498.

Warwick, J.W., 1989, Press Conference, JPL, August 27.

INTERNATIONAL COOPERATION IN MAGNETOSPHERIC PHYSICS

Juan G. Roederer

Geophysical Institute
University of Alaska Fairbanks
Fairbanks, AK 99775-0800

ABSTRACT

International scientific cooperation in the geosciences started long ago. The main driving force was the need to make measurements simultaneously at different points of the globe. Today, especially in magnetospheric physics, there are additional reasons, one of which is the need to marshal human resources on a world-wide scale to study and understand the ever-increasing complexities of the system. The solution of "global" problems of magnetospheric physics calls for a new mission-oriented mode of work, based on a carefully planned community approach. An internationally coordinated interplay between experiment, theory and computer simulation studies is necessary, as well as a new interactive mode of sharing information and data. In this age of endemically declining budgets for basic science, the international community of magnetospheric physicists must join in an all-out effort to chart their work according to consensus priorities and it must devise strategies to transmit the excitement of their science effectively to the world public.

Much has been written and said about the role of international cooperation in science. About the fact that science is the only real *common* language of mankind - a basis for understanding and communication of ideas between people of different cultures and ideologies. About the fact that international coordination of scientific research brings together not only people from different political systems, but also people from different areas of economic development.

The geosciences have always been at the forefront of international scientific cooperation. This started long ago at the time when seafarers exchanged information on the oceans, their currents and prevailing winds. Today, geophysical research is becoming complex and costly, focussing on global studies conducted simultaneously with standardized instruments distributed over the whole planet; requiring costly expeditions; observatory chains; multiple satellite missions; supercomputer centers; and sophisticated data centers.

This demands international cooperation and the sharing of financial resources. But above all, it requires the marshaling of human resources - a sharing of intellect and expertise to understand the ever-increasing complexities of our environment that are being revealed by

Magnetospheric Physics, Edited by B. Hultqvist and C.-G. Fälthammar
Plenum Press, New York, 1990

better and more precise instruments and by more complete and densely distributed observatories.

The study of the magnetosphere (which really began when Jim Van Allen demonstrated that way out there, beyond the ionosphere, there was something *physically* tied to our planet - magnetically trapped particles) is a prime example of a new discipline revealing a constantly increasing complexity with a constantly increasing dependence on international cooperation.

Magnetospheric physics has undergone several stages of development - from a phase of discovery, to a phase of synoptic description of major regions and dynamic phenomena, to the current phase of trying to understand the magnetosphere's global behavior and response. International cooperative programs such as the International Magnetospheric Study 1976-79 (IMS) organized by the Scientific Committee on Solar-Terrestrial Physics (SCOSTEP) have played an important role in making these transitions possible.

I believe that yet another kind of transition is necessary. I feel that our field still is fragmented into too many individual "pet projects", each one with too narrow a perspective. We are still too preoccupied with oversimplified, cartoon-like, mostly two-dimensional, mostly quasisteady views of the real world, which of course is four-dimensional, eminently non-linear, topologically complex and difficult to visualize, chaotic, dominated by instabilities, discontinuities and catastrophic transitions, and composed of subsystems that usually lack ground state or dynamic equilibria. Yet in this chaos there *is* order and structure determining many macroscopic properties of the overall system. Our challenge now is to find ways to identify this order, describe it, and make use of it for both scientific understanding and practical prediction-making.

The understanding of the magnetopause and adjacent boundary layers is a case in point. I am sure that if we could *see* the boundary in its entirety, we would be horrified at its ugliness - and discouraged by its dynamic and chaotic complexity, even for the case of a constant solar wind flow! Like in so many biological systems, the complexity is so great that probably even model or theory so far formulated for the boundary may be correct to a certain extent - what needs to be determined is their *relative* importance or frequency of occurrence.

There is only one way to attack the formidable task of achieving a comprehensive understanding of the magnetopause: it calls for a new "mission-oriented" mode of work based on a carefully planned cooperative, coordinated *community* approach, in which a limited set of agreed-upon preselected problems is addressed by different theoretical *and* experimental groups, working simultaneously but following different approaches, and comparing notes periodically in interactive workshops. We have to set up an international "cooperative thinking machine." In this new mode, an important task would be to revisit systematically past experimental data, especially from a space mission such as *HEOS*, which provided unique data on the distant cusp and the high latitude boundary layer.

I know that there are colleagues who still view the traditional "independent investigator" approach as the most desirable mode to advance theoretical understanding. No doubt that one mode should not replace the other. But collective creativity may be greater then the sum of its parts. Individual creativity cannot be learned and there are only very few truly independent creators in science. But *collective* creativity can be developed and stimulated and can lead to macroscopic results.

Coming back to our example of the magnetopause, clear and systematic questions should be posed and attacked. For instance: what kinds of ripples and waves are caused by which irregularities in the solar wind? What about mesoscale features on the boundary such as filaments, "balloons" and "blisters"? What features in the solar wind trigger flux transfer events (FTE); what, if any, FTE's are chaotic, of endemic origin? What is the 3-D macroscopic topological aspect of FTE's, how are they linked to reconnection sites? What is the fate of boundary layer flux tubes, how do they relate to current systems, what are their ionospheric footprints, what is the ionospheric feedback? Do FTE flux tubes quickly lose their identity inside the magnetosphere? How many types of reconnection processes are operating? What are the principal solar wind plasma entry mechanisms? Are reconnection events just mere manifestations of solar wind-magnetosphere coupling, or are they the principal cause of macroscopic reconfigurations? Have we exhausted the study of all possible classes of reconnection, or does turbulent reconnection offer new possibilities of understanding?

There are just a few of the many questions that await an answer in just *one* area of space plasma physics! In spite of this proliferation of questions, we must be selective and we must have the guts to agree on priorities! As one colleague put it at a recent meeting, "one has to be careful that our field does not become like philosophy, which never provides any answers but in which the questions are continuously getting better!"

On the other hand, almost paradoxically, it is not possible anymore to concentrate on just one limited area while closing the eyes to the grand picture and how one's own area fits into it. This grand picture relates to what I like to call "the ultimate questions." For instance, why is it that whenever Nature makes plasma, She arranges it into discrete regions separated by thin boundaries or discontinuities which exert an active control of the dynamics of the regions adjacent to them (Hannes Alfvén's cellular plasma universe)? Are field-aligned electric fields a universal feature whenever a collisionless plasma region is magnetically coupled to a resistive region? Is the magnetospheric substorm the manifestation of a universal global plasma instability?

In space plasma physics it is becoming more evident than ever that progress will depend on the close interplay between experiment, theory and simulation. The latter research mode plays a crucial role because we are dealing with a system in which nothing is proportional to anything and everything is coupled to something else! Of the three modes of research, simulation is perhaps the most difficult one to do *right*. By "right" I mean nontrivial. Sometimes simulation studies only produce pretty pictures; or they merely confirm that MHD or kinetic theory indeed do work well - or that they don't work at all because of too much numerical diffusion. A far less trivial task is to extract comprehensive, quantitative physical understanding for a simulation study. Verification of models is one way, systematic testing of parameters is another, but the perhaps most important application is that of identifying the basic component processes that occur at different scales (micro, meso, macro), their relative importance, and their mutual interaction - and use all this for quantitative prediction-making. I think that in space plasma physics we are still far from doing this on a routine bases.

One approach I would like to see as an international effort, a sort of an international contest, is a systematic, 3-D simulation study of "creating a magnetosphere" from scratch and examining in great detail the transient stages and processes that arise. For instance: 1) Begin with an unmagnetized solar wind, and for different dipole tilt angles, create an Axford-Hines configuration adiabatically; 2) Watch the formation of boundaries and plasma reservoirs for given plasma entry mechanisms and analyze the resulting convection patterns; 3) Plug in an ionosphere and determine the effects; 4) Examine the effects at impulsive variations of solar wind pressure, velocity and temperature at different spatial scale and watch how the tail evolves with distance under different spatial scales and watch how the tail evolves with distance under different stresses; 5) Turn on the interplanetary magnetic field (IMF) slowly, for different directions and re-examine the boundary, its inherent instabilities and particle entry processes under steady conditions; 6) Examine the evolution and instabilities of the plasma sheet, and the formation of ring current and radiation belts; 7) Vary the IMF impulsively and determine the effects; etc. etc. Follow the same steps in altered order to search for "hysteresis" effects (final state dependent on actual history of creation); compare the results of MHD, kinetic and hybrid codes, etc. Every time individual features or component processes are identified (such as boundary layers; driven or spontaneous reconnection sites; X-and O-line formation in the tail; plasmoids; field-aligned current sheets or filaments), look at them with a "magnifying glass" (i.e., with much higher resolution on much finer meshes and time steps) to determine specific features (e.g., formation of layered or filamentary structures in boundaries; the causal connection between Alfvén waves, field-aligned currents and parallel electric fields; feedback effects from the ionosphere). If done properly, this task of creating a magnetosphere would require a coordinated community effort with groups working independently, but toward agreed-upon common goals.

On the experimental side, the "new mode" of research cooperation would require more than ever the active sharing of data and the collective approach to coordinate data analysis. Within reasonable limits, the (mostly psychological) barriers of proprietary ownership of data, especially satellite data, have to be broken down to make data available to a maximum number of scientists. This has been done successfully in several recent missions (e.g., AMPTE) and I see no reason why it cannot be extended to future missions. The integration of space and ground-based observations is essential. Each class taken separately is a necessary but *not* a sufficient element of magnetospheric research.

These are just some of the "scientific realities" which propel us into cooperative modes of research. But there are also *political* realities. While the statements below are based mostly on my experience in the United States, they also may apply to other countries.

We live in the era of "big science," in the era of "global studies," in the era of "centers of excellence." But we also live in an era of declining or stationary budgets, increasing public distrust in the scientific establishment, and general lack of public understanding of the aims of science. All this forces the funding agencies to concentrate on a limited number of big projects of "societal relevance" or with "regional economic spinoffs," to the detriment of basic research and the support of individual investigators.

Scientific fraud in the biomedical sciences, predictions by unscrupulous scientists using still oversimplified climate models and questionable data to exploit the global warming scare, power-hungry scientists and money-hungry university administrators taking advantage of the scientific illiteracy of the media to announce to the world phony achievements such as "cold fusion," are all responsible for extending the public distrust to *all* disciplines.

As a result, it is not anymore enough to submit "good" research proposals: they have to fit into the plan and priorities of the funding agency. It is not anymore enough for an investigator to fight for his/her own money: investigators must first band together and fight for money allocations *within* a given agency to their own field of interest. This requires an even *earlier* effort of first selling their field to colleagues in other disciplines. Even if financial appropriations to a given agency increase, the budgets of individual programs may decline through internal reallocations if such programs are of low priority. Within an agency, there are now only two realistic alternatives to get a substantial increase for a given program. One is to develop a program of *national* scope; the other is to take money away from other programs or projects that have lower priority ranking within the agency. Either way requires community support and involvement to succeed.

In countries where peer review is used to screen research proposals, this process has turned "bloody", especially in areas in which those peers have not banded together to first fight jointly for overall increased funding for their own field. Gone is the flexibility of program directors to ignore unfair proposal reviews or to promote a subject that is somewhat out of the line of current thinking and thus bound to receive lower marks in peer review: a proposal is only as good as the *worst* review it gets!

Unfortunately, the younger generation of scientists (and a good many old-timers too!) has not yet fully comprehended that unless they participate themselves in the process of planning, prioritizing, and related compromising, somebody else will do this for them and their chances of receiving adequate funding will continue to decrease - no matter how good their individual proposals are. And many universities have not comprehended this either, and do not adequately support their scientists in these "public policy" endeavors.

Magnetospheric physics has not escaped these problems. It involves a relatively small and partly fragmented community compared for example to astronomers. The astronomers are fragmented, too, but when it comes to defending their field before funding agencies and the public, they act in highly organized fashion. And the global warming issue presents itself as a real threat, not only to our planet, but also to the fiscal health of solar-terrestrial research. This latter situation is particularly worrisome for ground-based, theoretical and modeling/simulation studies of the magnetosphere and upper atmosphere. While some aspects of middle atmosphere aeronomy are slowly, but successfully, being incorporated into the international Global Change program (which has an eminently biospheric focus) and while space observations of the magnetosphere have a good future in the satellite missions of ISTP or STSP, the ground-based magnetospheric programs as well as related theoretical and simulation studies are left to compete with the Global Change program which in the meantime has reached household-word status.

This is where the Solar-Terrestrial Energy Program (STEP) comes in. SCOSTEP is in the final stages of planning the biggest international cooperative program ever attempted. The main goal of STEP is to advance the quantitative understanding of the coupling mechanisms that are responsible for the transfer of energy and mass from one region of the solar-terrestrial system to another. The magnetosphere will play, of course, a fundamental role in this program. STEP will involve ground-based, aircraft, balloon, rocket and satellite experiments; theory and simulation studies; and dedicated data and information systems. Integral to the success of STEP is a set of 13 solar-terrestrial spacecraft missions approved by the Inter-Agency Consultative Group as the next cooperative project of NASA, ESA, ISAS, and INTERCOSMOS.

Let me just point out one important potential side-benefit of STEP: the promotion of public understanding of our discipline. To achieve this goal we must learn to talk to the public and convey to them the results of *our* science in *their* language. This is not easy for magnetospheric physics. But I believe it can be done. The promotion of concepts like "climate and weather in space" will be helpful. The concept of the polar upper atmosphere as our "window to outer space" is helpful. Comparing the cellular structure of the magnetosphere with that of the lithosphere (rocks, liniments, strata, plates separated by cracks and faults), and describing substorms as "plasma quakes" has proven effective to explain this phenomenon to colleagues from the solid earth sciences and to the public at large. Describing auroral displays as "space shows" projected on the giant TV screen of the upper atmosphere, and visualizing the actual "astronomical" size of the unfortunately invisible planetary magnetospheres have been effective tools to promote the public understanding of our science.

To sum up, in addition to joining in an all-out effort to unravel the outstanding problems of solar-terrestrial research, we must unite in an all-out effort of bringing our science to the world public. Let us make the magnetosphere a household word, too!

THE NEED FOR HIGH TIME RESOLUTION MEASUREMENTS IN THE MAGNETOSPHERE

F. S. Mozer

Physics Department and Space Sciences Laboratory
University of California
Berkeley, California

ABSTRACT

Magnetospheric phenomena are characterized by the formation of thin boundaries, across which energy is converted from one form to another. These boundaries and energy conversion processes are controlled by non-linear microphysics that occur on time scales as fast as 1 to 100 microseconds. The understanding of these phenomena requires measurements of plasma distributions and time domain electric fields with appropriately high time resolutions. Several examples of such measurements are given and some planned satellite experiments are described.

1. INTRODUCTION

In addition to honoring a pioneer in our field, the Crafoord Symposium offers us the opportunity of standing back from our research to consider which branches of magnetospheric physics offer the best prospects of producing important scientific results in coming years. This topic may be approached by noting that all fields of physics evolve through stages, which may be classified as:

a. Discovery

b. Description

c. Interpretation in terms of fundamental physics.

The first of these phases for terrestrial magnetospheric research ended about a decade ago with the discovery of the auroral acceleration region. The era of description can last forever because the magnetosphere is three-dimensional, time varying, and incredibly complex. As future research programs are planned and new instruments are developed, we must be wary of investing too much effort in the description of phenomena at the expense of concentrating on the fundamental physics of these phenomena. For example, one could describe the clouds in the sky forever because they are beautiful and complex, and they change all the time. However, a prudent person would stop emphasizing descriptions after a time and would start concentrating on the physics of condensation, wind shear, electrification, etc.

Magnetospheric Physics, Edited by B. Hultqvist and C.-G. Fälthammar
Plenum Press, New York, 1990

Magnetospheric research must emphasize quantitative science over description even though little more than cartoons of many of its features exist. The interesting physics of the magnetosphere is associated with its ability to develop thin boundaries between different regions, with energy converted from one form to another (kinetic energy of flow, Poynting flux, particle thermal energy, etc.) across them. Thus, magnetospheric research should focus on the physics of the formation and maintenance of these boundaries and on the energy conversion that occurs across them. These processes all depend on non-linear microphysics occurring on short spatial scales and on time scales as fast as the plasma period, which ranges from one to 100 microseconds. Thus, new missions and detectors must emphasize plasma distribution function and time-domain electric field measurements on appropriately fast time scales. The purposes of this paper are to present examples of the need for such high time resolution measurements and to describe some satellite missions that will provide them.

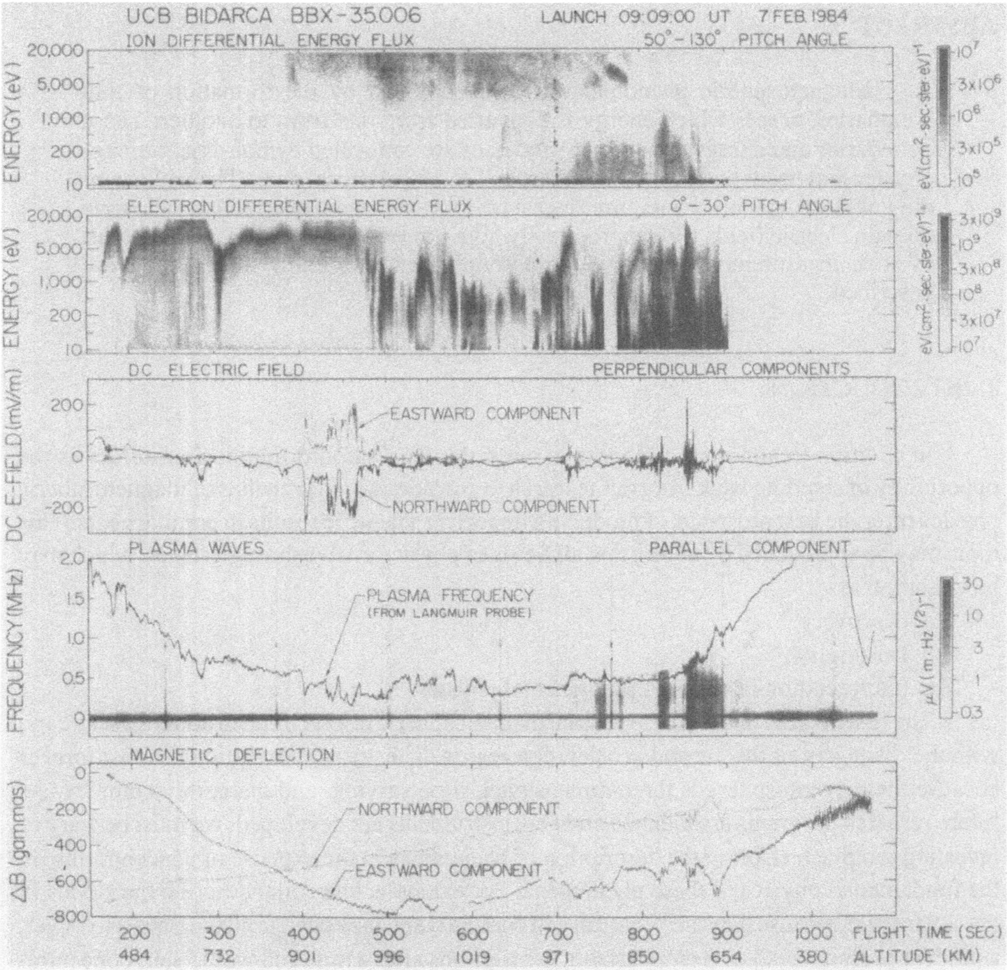

Fig. 1. Data from a complete sounding rocket flight (1100 seconds) through an intense auroral situation (after Boehm et al., 1989).

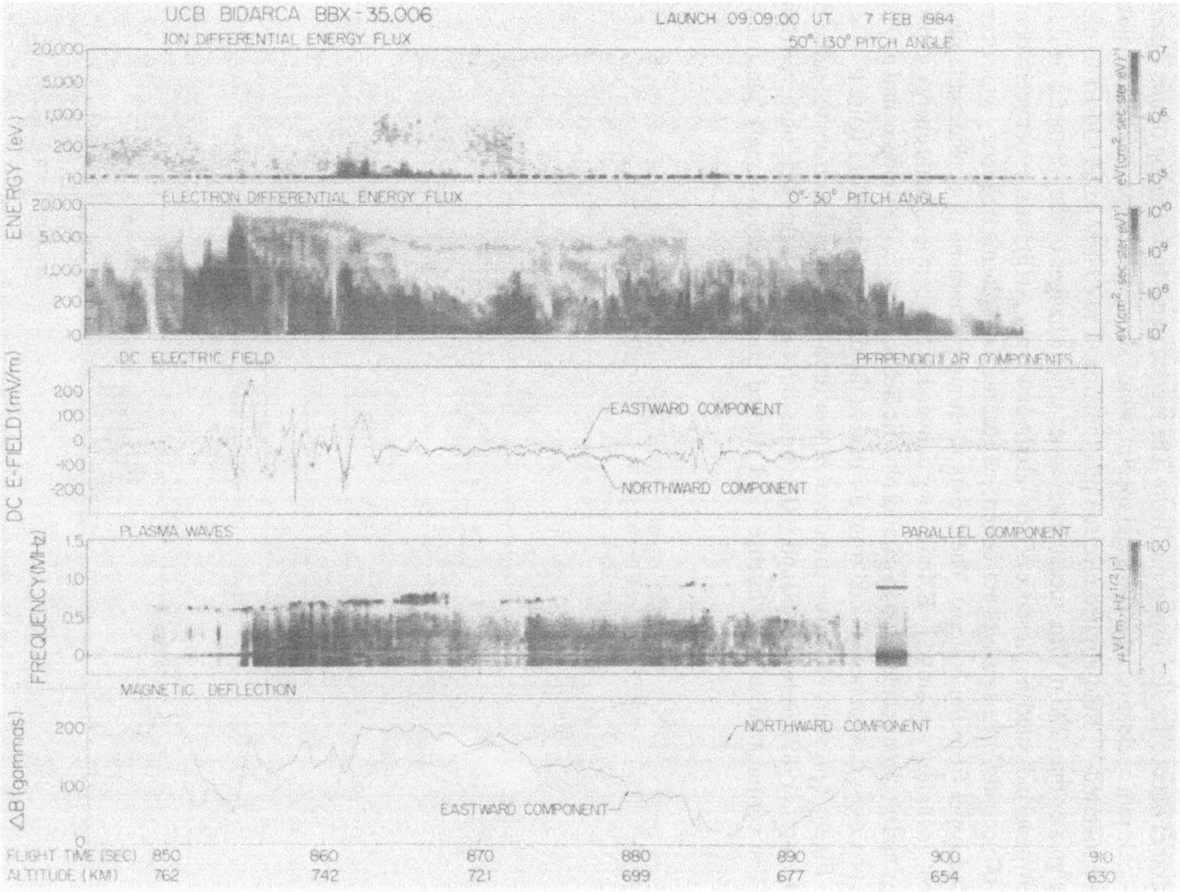

Fig. 2. Data for 65 seconds from the same rocket flight as in Figure 1 (after Boehm et al., 1989).

2. A FIRST AURORAL PHYSICS EXAMPLE

As the first example, data from the Berkeley sounding rocket program are presented because they possess the highest time resolution obtained to date in the magnetosphere. In Figure 1, sounding rocket measurements obtained on a flight from Alaska into an intense aurora are given (Boehm et al., 1989). These data have a time resolution equivalent to that of typical satellite measurements in the 1980s, with the time span of Figure 1 being the fifteen-minute duration of data collection during the flight. The second panel from the top of this figure is a gray-scale plot of electron intensities and spectra, which shows that the payload passed through intense inverted-V, field-aligned electron precipitation. The central panel illustrates electric fields greater than 200 mV/m, measured at times when changes of the magnetic field in the bottom panel indicated the presence of field-aligned currents. At flight times between about 800 and 900 seconds, intense waves at frequencies near the local plasma frequency were observed (second panel from the bottom). These data illustrate phenomena of considerable physical interest, including a positive **j • E**, indicating conversion of electromagnetic energy to particle energy and electron acceleration in parallel electric fields. Even so, the fundamental physics of these processes cannot be deduced from the data at the time resolution of Figure 1.

Figure 2 presents about one minute of higher time resolution data from the same flight (Boehm et al., 1989). Between 850 and 860 seconds, large electric fields were observed (middle panel) in conjunction with an intense field-aligned current (bottom panel), plasma wave

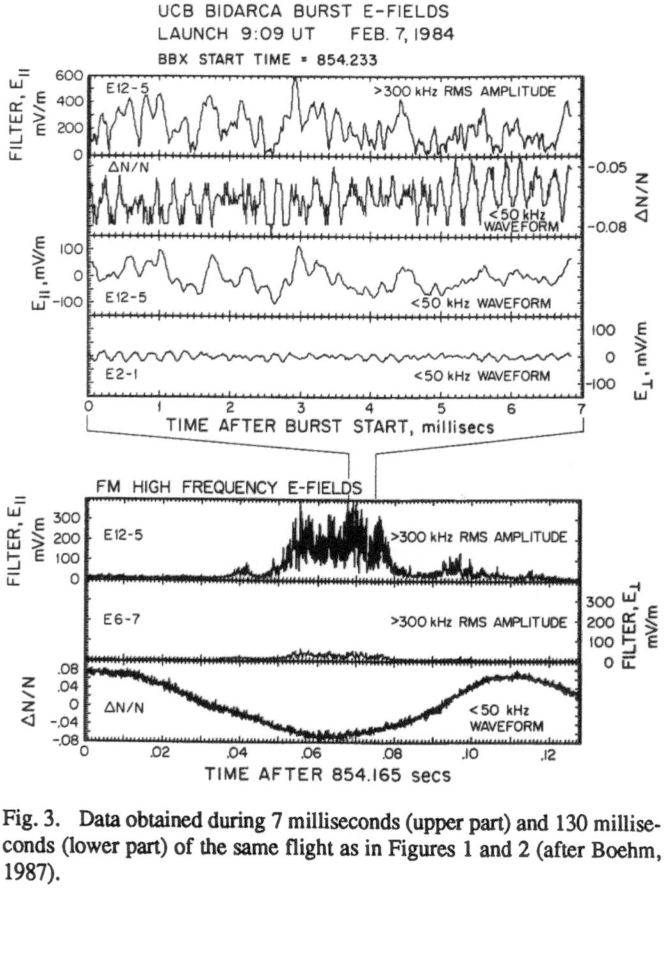

Fig. 3. Data obtained during 7 milliseconds (upper part) and 130 milliseconds (lower part) of the same flight as in Figures 1 and 2 (after Boehm, 1987).

emission (second panel from the bottom), and an increase in the electron flux and energy (second panel). At this time resolution the data are displayed in more detail, but they still cannot be analyzed quantitatively to determine fundamental physics.

The time interval near 854 seconds is examined with still higher time resolution in Figure 3 (Boehm, 1987), the bottom three panels of which cover an interval of 130 milliseconds. The uppermost of these panels gives the amplitude of the field-aligned component of the wave electric field near the plasma frequency. For about 20 milliseconds near the center of this figure, the plasma wave intensity exceeded several hundred mV/m, during which time the plasma density decreased (bottom panel). The central seven milliseconds of these data are expanded and illustrated in the top group of four panels. The first of these panels is a plot of the amplitude of waves near the plasma frequency, which shows several bursts of Langmuir waves having durations of a small fraction of a millisecond and amplitudes larger than 0.5 V/m. And thus, through improving the time resolution of data collection on sounding rockets to better than 100 microseconds, large amplitude Langmuir solutions have been discovered in the auroral ionosphere. These structures must be pervasive since they have been found on each of five rockets that was instrumented to look for them. They must also be important to auroral dynamics because of their large amplitudes.

To understand the source and interactions of these waves, even higher time resolution measurements were required. In Figure 4, 10 seconds of data from a later rocket flight that carried new and faster detectors are illustrated (Ergun et al., 1989). Again, pulses of Langmuir waves with amplitudes greater than 100 mV/m were observed (second panel from the bottom). On this flight the frequency of the high frequency waves was measured by counting zero crossings of the raw electric field waveform. For broadband waves, this frequency should be variable and difficult to interpret, while, during periods of pure sine waves, this frequency should be constant at the wave frequency. The bottom panel of Figure 4 is a plot of this zero crossing frequency, which, during the large amplitude events, was constant at the local plasma frequency, indicating that pure Langmuir waves were being observed. As the plots in the top panel of Figure 4 illustrate, these waves were observed at times of large and changing electron fluxes.

Better than one microsecond resolution was achieved on this flight with a wave-particle correlator that tallied counting rates in each half cycle of the Langmuir wave to measure the existence and energies of electrons trapped in the wave. In Figure 5, the electron fluxes are displayed for a time of one second along with horizontal bars that give the times when electrons of different energies were observed by the wave-particle correlator to be trapped in the Langmuir wave (Ergun et al., 1989). The physical interpretation of these data is that electrons were impulsively accelerated to energies of several keV at an altitude of about 8000 kilometers, after which they experienced velocity dispersion in reaching the rocket altitude, as may be seen in Figure 4. This dispersion created unstable distribution functions at the rocket, with the Langmuir waves being energized by those electrons in the unstable part of the distribution. And thus, rocket measurements having time resolutions of less than a microsecond have revealed a previously unexpected feature of the auroral zone and have allowed its quantitative interpretation in terms of fundamental plasma physics.

3. OTHER AURORAL PHYSICS EXAMPLES

As the next example of high time resolution measurements, data obtained in the auroral acceleration region will be described. This region exists at an altitude of about 8000 kilometers, where auroral electrons such as those of the previous figures are accelerated. Measurements on

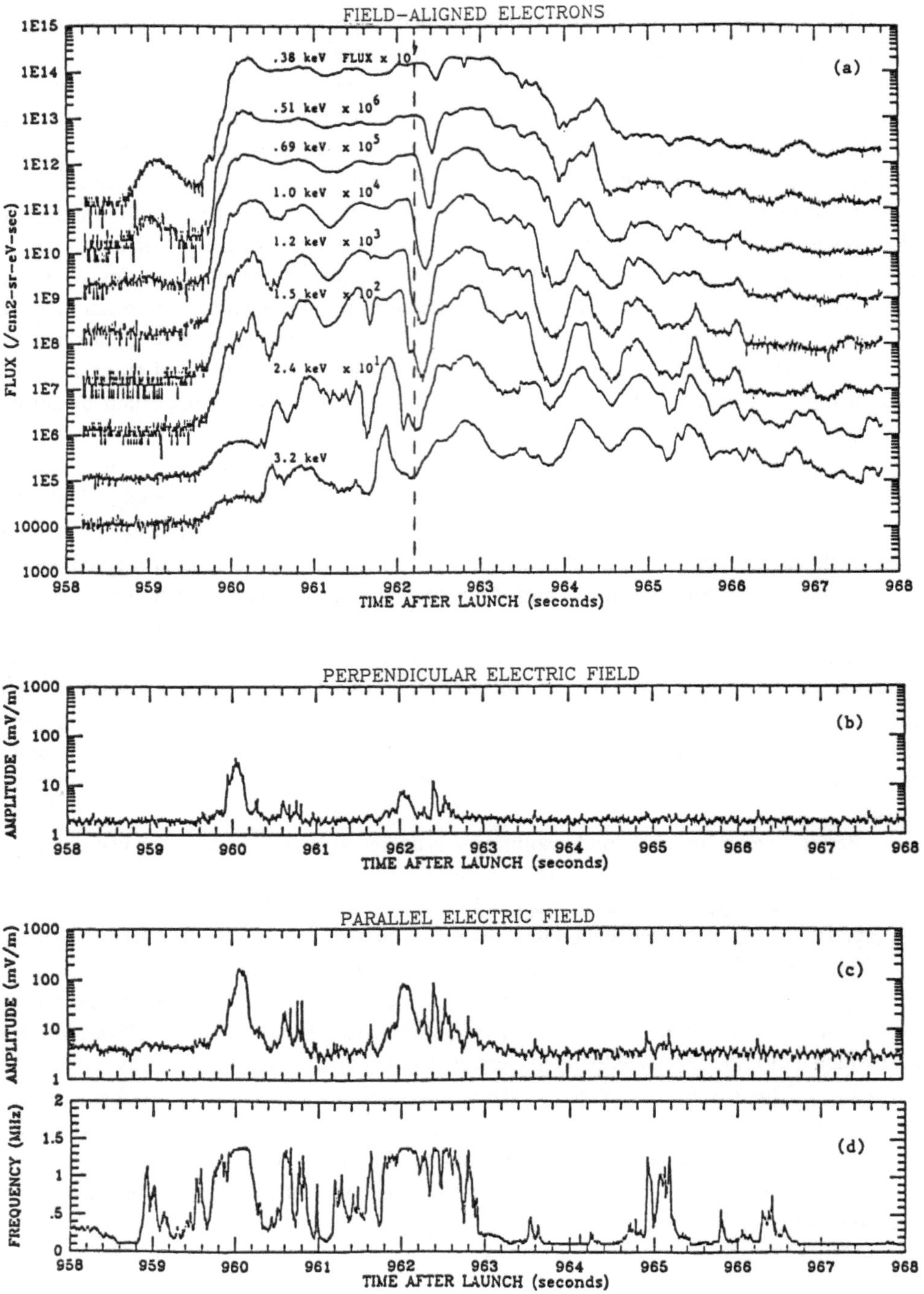

Fig. 4. Ten seconds of particle and field data from another auroral rocket flight that carried faster detectors than the one illustrated in the previous figures (after Ergun et al., 1989).

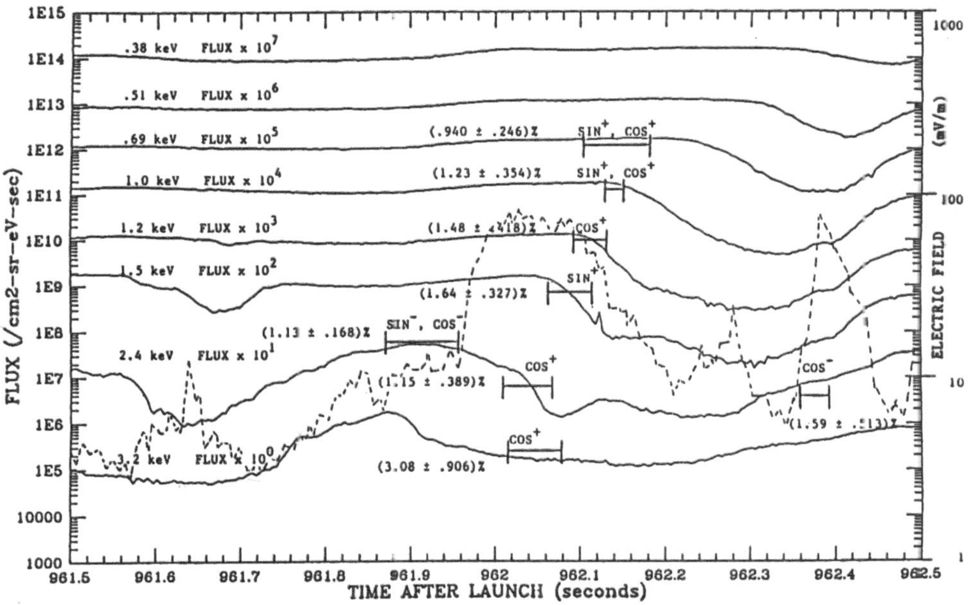

Fig. 5. One second of field-aligned electron and electric field data from the same flight as in Figure 4 (after Ergun et al., 1989).

S3-3 and Viking have revealed two types of time-domain electric field structures which are largely responsible for this acceleration (Mozer et al., 1977; Block et al., 1987; and Temerin et al., 1982). Examples of the first of these structures, called electrostatic shocks, are illustrated in Figures 6 and 7 (Temerin et al., 1981 and Block et al., 1987, respectively). These structures contain quasi-static, nearly perpendicular, electric fields as large as 500 mV/m that have durations in the satellite frame of typically 0.1 to 10 seconds. The second of the electric field structures discovered in the auroral acceleration region, called double layers (Temerin et al., 1982; Holback et al., 1986), is illustrated in Figure 8 (Mozer and Temerin, 1983) as the several millisecond duration parallel electric field pulses (bottom trace), which typically consist of a smaller positive (downward) field followed by as larger negative (upward) field component such that there is a net upward potential of a few volts across each double layer.

Before considering examples of how better time resolution measurements will clarify the physics of auroral particle acceleration in these field structures, several possible misconcep-

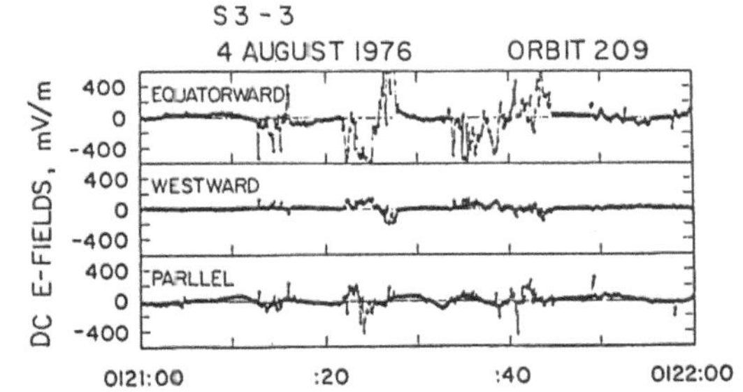

Fig. 6. Observations of electrostatic shocks on the S3-3 satellite (after Temerin et al., 1981).

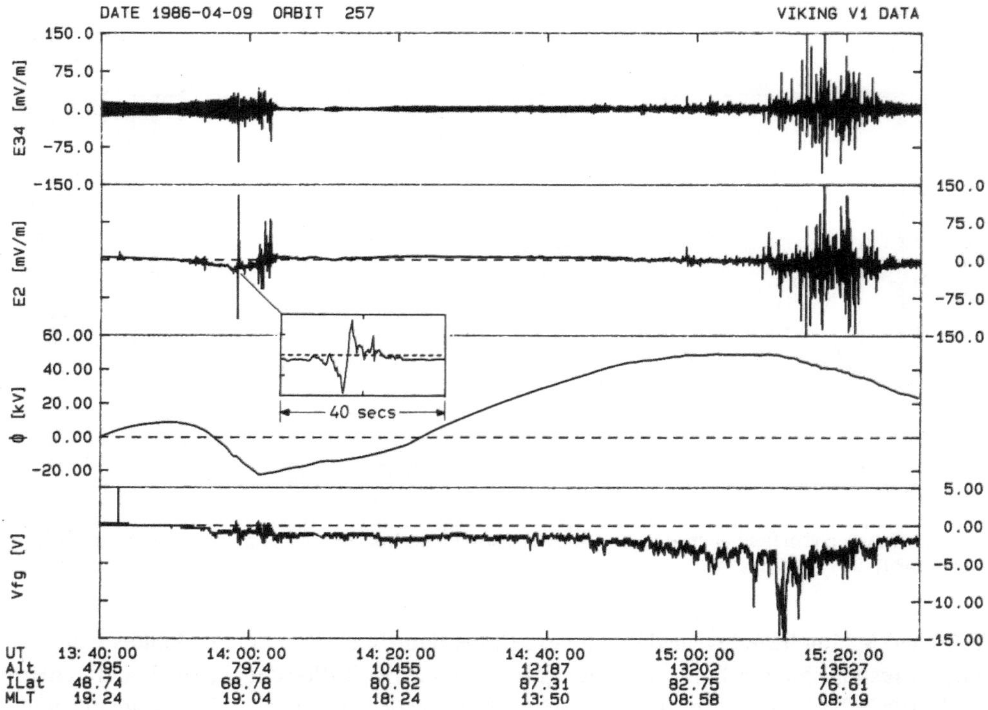

Fig. 7. Observations of electrostatic shocks on the Viking satellite (after Block et al., 1987).

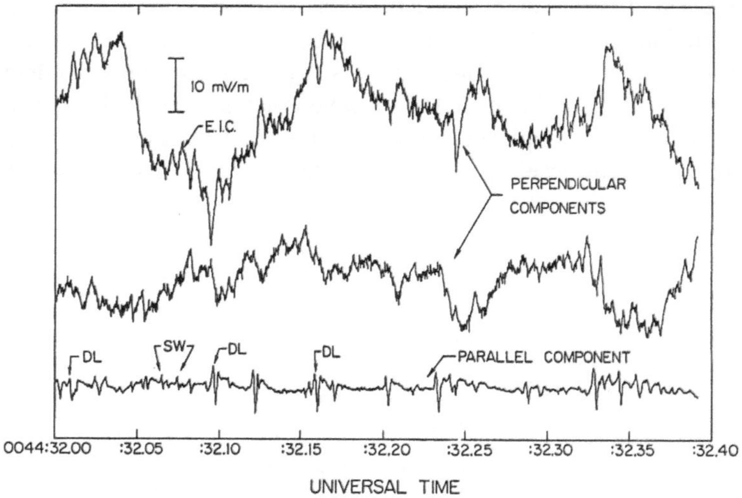

Fig. 8. Double layers (DL) observed with the S3-3 satellite (after Mozer and Temerin, 1983).

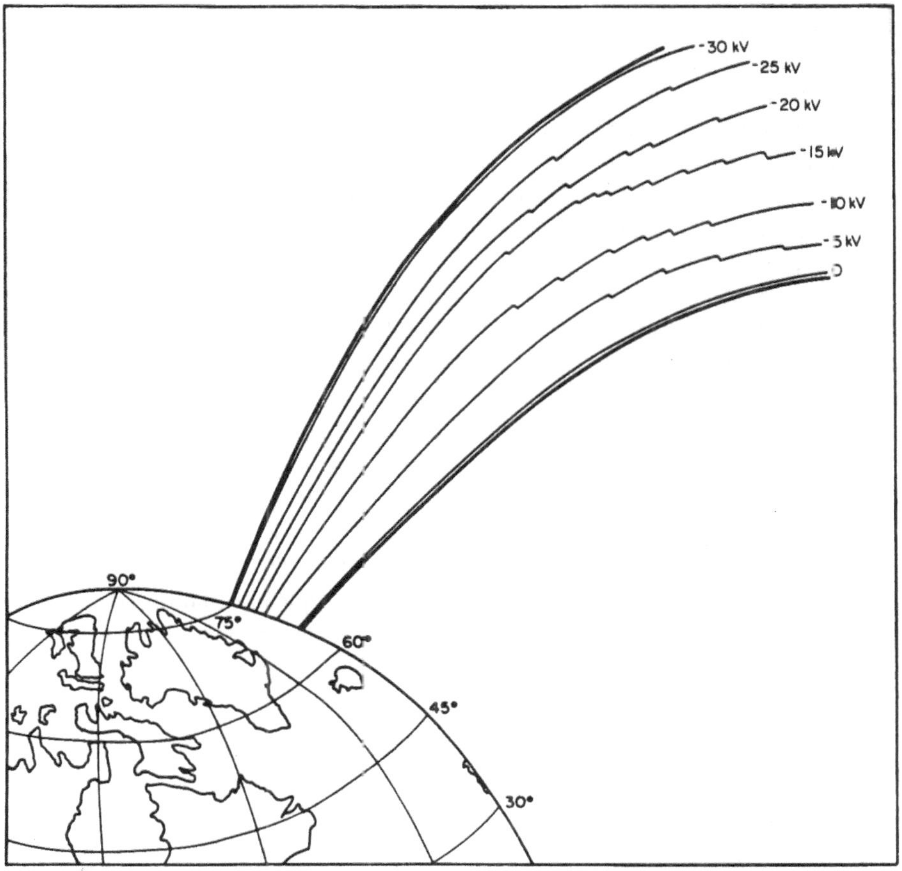

Fig. 9. Equipotential lines in the auroral region (after Mozer and Temerin, 1983).

tions about electrostatic shocks and double layers that have arisen as a result of oversimplified models will be discussed. These misconceptions include:

1. Electrostatic shocks always contain paired, oppositely directed, electric fields. Thus, the potential structure associated with them is always U-shaped. Paired, oppositely directed electric fields and the inferred U-shaped equipotential structures are observed in Figures 6 and 7 and in about 35% of the observations (Mozer et al., 1980). Generally, the electric fields in shocks are dominantly of one polarity, so their potentials are S-shaped. Sometimes the equipotentials deduced from the data are quite complicated (Temerin et al., 1981).
2. The parallel electric field in electrostatic shocks is contained in double layers. This statement is incorrect, at least some of the time, because non-zero, non-double layer, parallel electric fields have been measured in some shocks (Mozer et al., 1980; Block et al., 1987). This statement is unlikely to be correct at other times since shocks and double layers are rarely, if ever, observed in the same auroral structures.
3. Most of the parallel particle acceleration is caused by double layers.

Double layers occur less frequently than electrostatic shocks, and the energies of ion beams associated with double layers are generally much less than one keV, while several keV beams are often seen in electrostatic shocks. Although these results could be due to biases in spatial sampling, it is more likely that double layers are less important than electrostatic shocks in producing auroral particle acceleration.

As an example of fundamental physics that will be learned through higher time resolution measurements in the auroral acceleration region, particle acceleration in double layers will be considered. Through computer simulations and theory, the physics of single, isolated double layers is well-understood. However, since the total potential in a single double layer is less than 10 volts, they can only exert a significant influence on auroral particles if hundreds of them exist on a given magnetic field line at the same time, as illustrated in the model of Figure 9 (Mozer and Temerin, 1983). If such is the case, then each double layer modifies the electron distribution that is incident on the neighboring double layers in a way that requires experimental data for its full understanding. Thus, electron distribution functions must be measured with a time resolution of better than one millisecond in the auroral acceleration region before the processes of particle acceleration in double layers can be fully understood. The highest resolution data to date fail this requirement by more than three orders of magnitude.

As another example of the need for high time resolution data in the auroral acceleration region, ion acceleration in electrostatic shocks will be considered. Since the dimensions of electrostatic shocks are comparable to the ion gyroradius, the interaction of ions with shocks can involve processes that occur in a fraction of a gyroperiod. Thus, complete ion distributions must be measured in a time that is short compared to the typically 10 millisecond proton gyroperiod. This requirement has not been satisfied in measurements to date by at least three orders of magnitude.

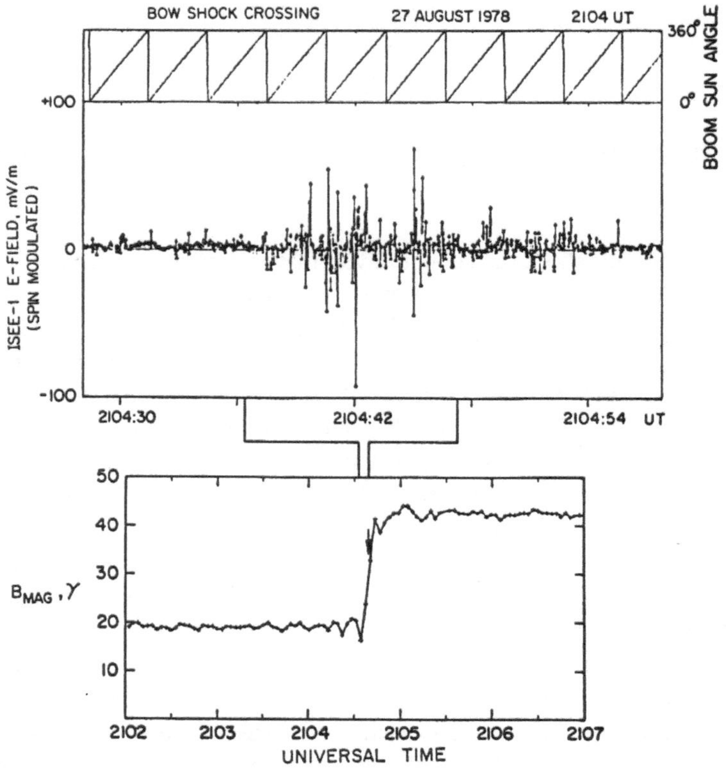

Fig. 10. Single component, raw electric field and magnetic field magnitude data from bow shock crossing by the ISEE-1 satellite (after Wygant et al., 1987).

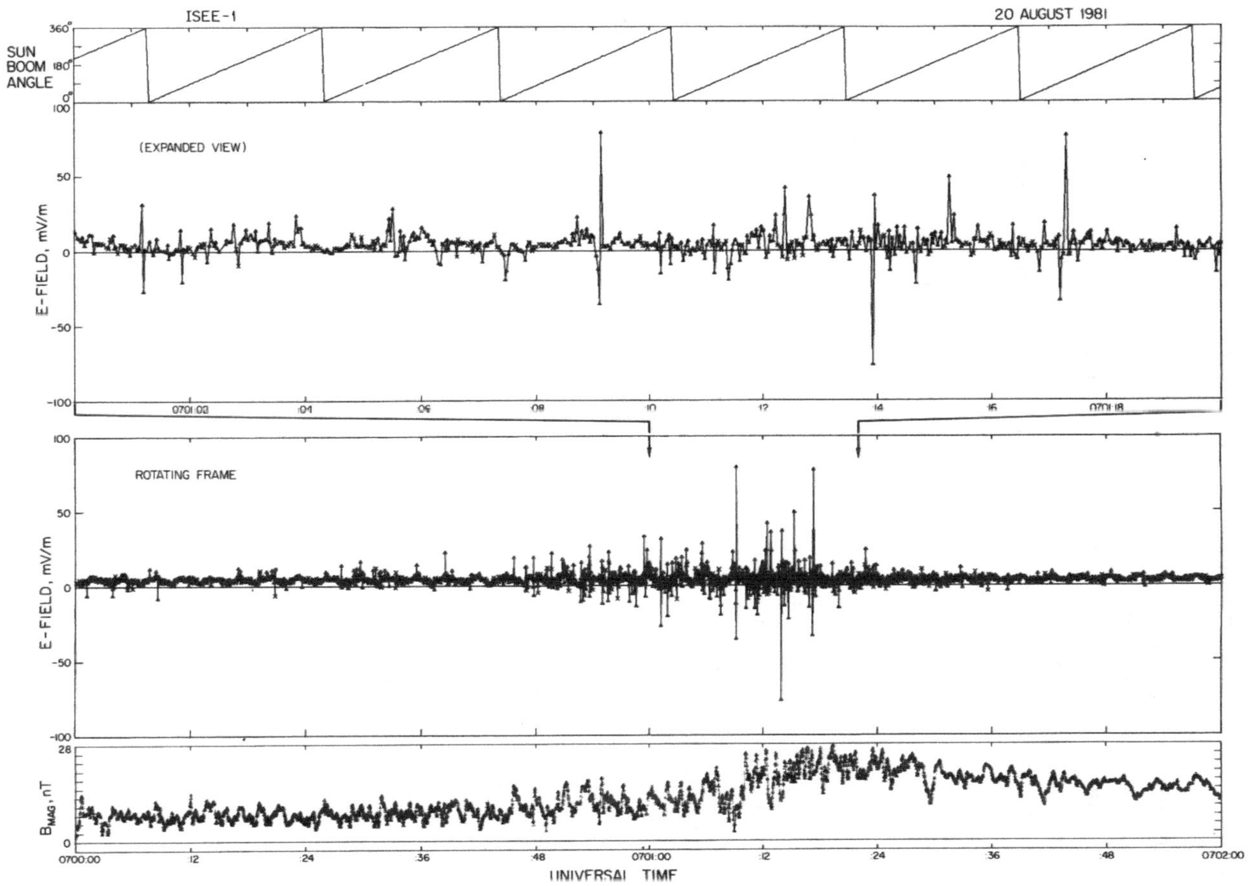

Fig. 11. Single component, raw electric field data and magnetic field magnitude data from another ISEE-1 bow shock crossing (after Wygant et al., 1987).

Before considering other examples of the need for high time resolution measurements in the magnetosphere, an important point will be emphasized. The interesting physics in the examples thus far presented has been obtained by considering electric field measurements plotted as a function of time, i.e., from time-domain analyses of electric field data. As W. Lotko (private communication) has shown, spectra of the electric fields in double layers (i.e., frequency-domain presentations) are rather nondescript and do not reveal the interesting physics that is present in the data. This is because the nonlinear physics at magnetospheric boundaries causes the electric fields to clump in space and time in a way that is best displayed by time-domain presentations. On the other hand, linear processes in plasmas, involving resonances and cut-offs, are best displayed in the frequency domain. The implication of these considerations is that the high time resolution electric field measurements required for studying the fundamental, nonlinear, physics of magnetospheric boundaries must be obtained in the time domain, that is, through collection and display of electric field components as a function of time.

4. EXAMPLES FROM OTHER REGIONS OF THE MAGNETOSPHERE

The highest time resolution data that have been obtained to date in other regions of the magnetosphere are single-component electric field measurements made on ISEE-1 with a time resolution of about 0.1 seconds. Although such data are insufficient for quantitative studies, they do support the view that relatively unobserved processes remain to be studied, understood, and incorporated into the physics of magnetospheric boundaries. The following discussions illustrate these points.

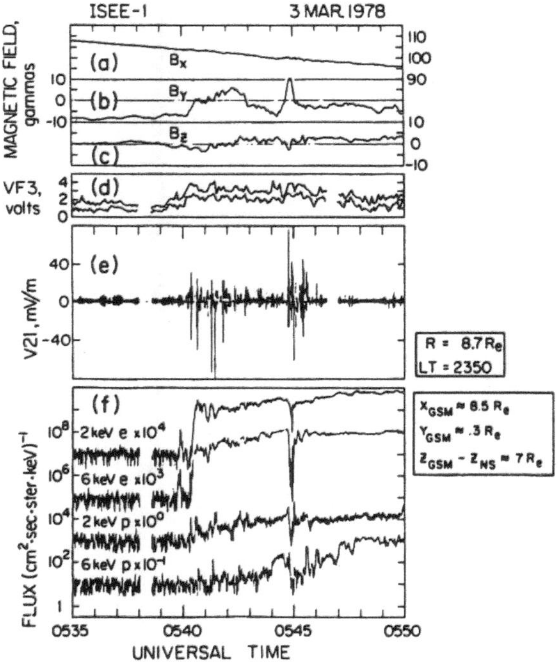

Fig. 12. ISEE-1 data obtained at the high latitude boundary of the plasma sheet in the geomagnetic tail (after Cattell et al., 1982).

The bottom panels of Figures 10 and 11 present ISEE-1 measurements of the total magnetic field at two different bow shock crossings (Wygant et al., 1987). Raw electric field data obtained during the magnetic ramps are illustrated in the center panels of each figure. These single-component electric field measurements show several examples of unresolved electric field variations as large as 100 mV/m, which often appear to be structures containing opposite polarity components. Since the double layers in the auroral acceleration region have such an electric field structure, the data of Figures 10 and 11 suggest the possibility that double layers or similar structures exist in the bow shock. Clearly, present models of particle heating and acceleration at the bow shock will be greatly modified when information at the required high time resolution becomes available to incorporate these large electric fields into collisionless shock theory.

Such large, impulsive electric field structures are also observed at the high latitude boundary of the plasma sheet in the geomagnetic tail (Cattell et al., 1982), as the data of Figure 12 illustrate. The ion and electron fluxes in the bottom panel of this figure increase near 0540, indicating passage of ISEE-1 from the tail lobes into the plasma sheet. Near 0545, these fluxes decrease briefly, indicating that the spacecraft again encountered the high latitude boundary. At each of these encounters, the single-component electric field measurement (center panel) showed spiky structures with amplitudes approaching 100 mV/m. These large, spiky fields, which are seen on nearly every passage through the high latitude boundary of the plasma sheet (Levin et al., 1983), must also exert a significant influence on the physics of this region. This influence will only be understood when adequate high time resolution measurements are made.

The last example of very large, short duration, electric fields occurred at the neutral sheet in the geomagnetic tail during the passage of a near-earth plasmoid over ISEE-1 (Nishida et al., 1983). The spacecraft was near the neutral sheet during most of the 90 seconds illustrated in Figure 13, although it was probably closest during the interval between the two vertical lines in this figure because B_x (second panel from the bottom) was nearly zero. At this time, and also earlier, the measured electric field exhibited spiky structures having amplitudes larger than 200 mV/m. Again, nothing quantitative can be deduced from unresolved measurements of a single component of a three-component vector. However, it is clear that new fundamental physics of tail reconnection and plasmoid formation will be revealed when appropriately fast measurements of all components of the electric field and the plasma distributions are obtained near the neutral sheet.

In passing, it is worth noting that large, spiky electric fields like these illustrated in the previous figures have been looked for and not found at the dayside magnetopause and in flux transfer events, which, a priori, might be thought to be the most likely regions for finding such fields. The apparent absence of these structures at the magnetopause may be due either to inadequate time resolution in the measurements, or to the absence of such phenomena where they are most expected. In either case, the next generation of high time resolution measurements at the magnetopause will certainly be interesting.

5. FUTURE HIGH TIME RESOLUTION MISSIONS

Instruments for making higher time resolution measurements of plasmas and time-domain electric fields are being built for future missions. As an example, the electric field instrument on the NASA Polar satellite will make three-component measurements with a time resolution of less than 100 microseconds, through capturing the order of a minute of data in a two megabyte solid state memory. The electric field instruments on the ESA Cluster satellites will have similar time resolutions, but will measure only two of the three components of the

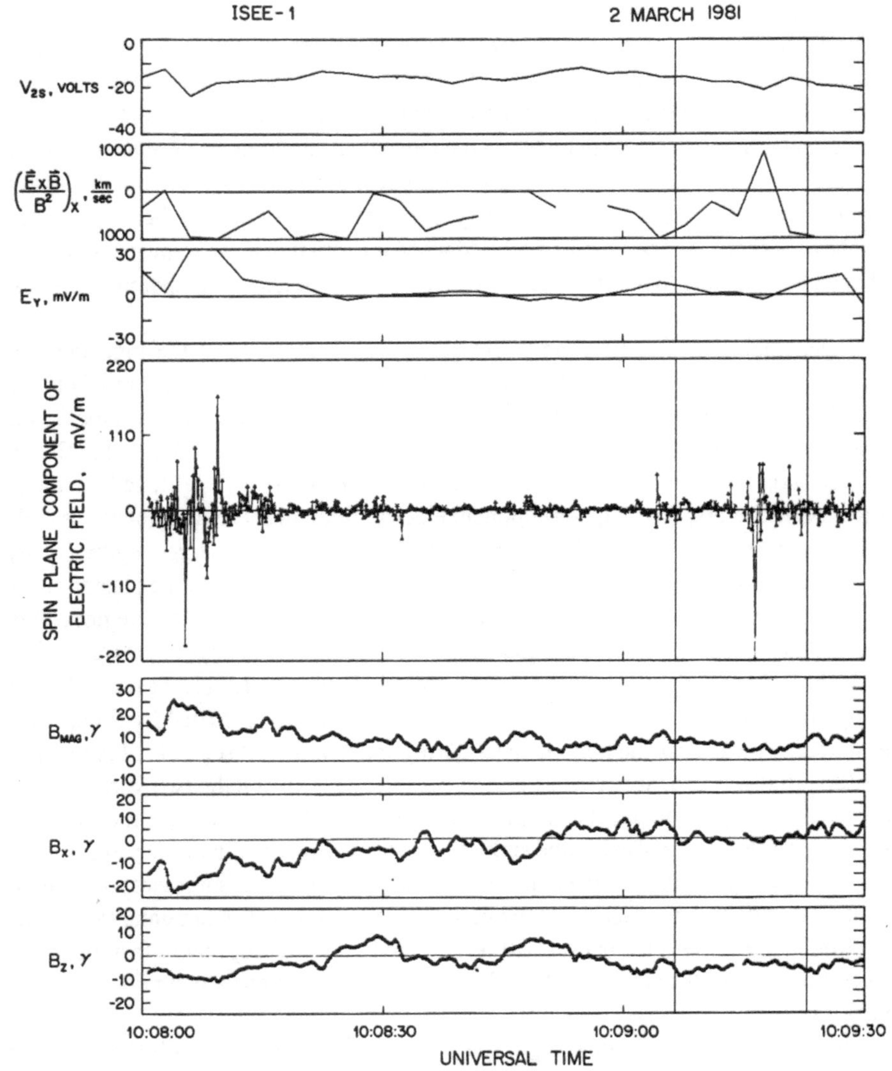

Fig. 13. ISEE-1 data from the neutral sheet in the geomagnetic tail during the passage of a plasmoid over the satellite (after Nishida et al., 1983).

electric field. Plasma instruments are being developed for both Polar and Cluster with burst memories for obtaining distributions with time resolutions of much less than one second.

The highest time resolution measurements on any spacecraft currently in fabrication will be made on the FAST Explorer satellite, which is part of the NASA Small Explorer Satellite program. Scheduled for launch in 1993, the instruments on FAST are being provided by the University of California, Berkeley; a Lockheed group which is developing the mass spectrometer; and a UCLA/Los Alamos group which is providing the search coil and fluxgate magnetometers. This satellite will be placed in a polar orbit with a 3500 kilometer apogee to study the physics of the lower auroral acceleration region. It will contain a 100 megabyte solid state memory that will collect data in ground-commanded formats and at data rates as high as 10 megabits/second for the order of a minute on high altitude auroral zone passes. Three-component, time-domain, electric field measurements will be made with time resolutions as fast as one microsecond. The boom-mounted spheres will also operate in a current mode to measure time-domain density fluctuations with a similar resolution.

The plasma detectors on FAST will also achieve the desired time resolutions. For example, the Fast Electron Spectrometer will measure 96 points on the electron distribution function with a time resolution of one millisecond. Two-dimensional ion distributions will be measured for both protons and oxygen with a 64 millisecond resolution, and selected pitch angles will be observed every two milliseconds.

The experiments to be flown on the FAST satellite illustrate that we will be able to measure plasmas and fields in the magnetosphere with the time resolution required to study the nonlinear microphysics of magnetospheric boundaries. Examples discussed in this article illustrate the possibilities for learning new and fundamental physics from such measurements. It remains a challenging task to organize future multi-satellite missions containing matched, high time resolution instruments that are designed in a complementary manner and that will work together in data collection and analysis to exploit high time resolution measurements.

REFERENCES

Boehm, M. H., 1987, Waves and static electric fields in the auroral acceleration region, Ph.D. dissertation, Physics Department, University of California, Berkeley.

Boehm, M. H., Carlson, C. W., McFadden, J. P., Clemmons, J. H., and Mozer, F. S., 1989, High resolution sounding rocket observations of large amplitude Alfvén waves, *J. Geophys. Res.*, in press.

Block, L. P., Fälthammar, C.-G., Lindqvist, P.-A., Marklund, G., Mozer, F. S., Pedersen, A., Potemra, T. A., and Zanetti, L. J., 1987, Electric field measurements on Viking: First results, *Geophys. Res. Lett.*, 14:435.

Cattell, C. A., Kim, M., Lin, R. P., and Mozer, F. S., 1982, Observations of large electric fields near the plasmasheet boundary by ISEE-1, *Geophys. Res. Lett.*, 9:539.

Ergun, R. E., Carlson, C. W., McFadden, J. P., and Clemmons, J. H., 1989, Langmuir wave growth and electron bunching: Results from a wave-particle correlator, *J. Geophys. Res.*, submitted.

Holback, B., Boström, R., Gustafsson, G., Koskinen, H., Holmgren, G., and Kintner, P., 1986, Propagating solitary plasma density structures observed by the Viking spacecraft, *EOS*, 67:1156.

Levin, S., Whitley, K., and Mozer, F. S., 1983, A statistical study of large electric field events in the Earth's magnetotail, *J. Geophys. Res.*, 88:7765.

Mozer, F. S., and Temerin, M., 1983, Solitary waves and double layers as the source of parallel electric fields in the auroral acceleration region, in: "High Latitude Space Plasma Physics," B. Hultqvist and T. Hagfors, eds., Plenum Publ. Corp., London, England.

Mozer, F. S., Carlson, C. W., Hudson, M. K., Torbert, R. B., Parady, B., Yatteau, J., and Kelley, M. C., 1977, Observations of paired electrostatic shocks in the polar magnetosphere, *Phys. Rev. Lett.*, 38:292.

Mozer, F. S., Cattell, C. A., Hudson, M. K., Lysak, R. L., Temerin, M., and Torbert, R. B., 1980, Satellite measurements and theories of low altitude auroral particle acceleration, *Space Sci. Rev.*, 27:155.

Nishida, A., Tulunay, Y. K., Mozer, F. S., Cattell, C. A., Hones, E. W., Jr., and Birn, J., 1983, Electric field evidence for tailward flow at substorm onset, *J. Geophys. Res.*, 88:9100.

Temerin, M., Cattell, C., Lysak, R., Hudson, M., Torbert, R. B., Mozer, F. S., Sharp, R. D., and Kintner, P. M., 1981, The small-scale structure of electrostatic shocks, *J. Geophys. Res.*, 86:11,278.

Temerin, M., Cerny, K., Lotko, W., and Mozer, F. S., 1982, Observations of double layers and solitary waves in the auroral plasma, *Phys. Rev. Lett.*, 48:1175.

Wygant, J. R., Bensadoun, M., and Mozer, F. S., 1987, Electric field measurements at subcritical, oblique bow shock crossings, *J. Geophys. Res.*, 92:11,109.

WHY WE NEED GLOBAL OBSERVATIONS

D. J. Williams

The Johns Hopkins University
Applied Physics Laboratory
Laurel, Maryland 20707
USA

ABSTRACT

Since its initiation with the remarkable discovery of the Van Allen radiation belts, the field of magnetospheric physics has been characterized by a continuing collection of data from *in-situ* satellite observing stations. The synthesis of these and ground-based observations has yielded an intriguing picture of a basic cosmological building block, the magnetosphere. Because of the time/space separation inherent in magnetospheric observations, this view necessarily is schematic in nature and our knowledge of global behavior is correspondingly incomplete. Since both local and global perspectives must be combined to understand such a complex physical system, the lack of appropriate global magnetospheric observations represents a serious impediment towards obtaining an understanding of magnetospheric dynamics. From the use of ground-based station networks to the advent of satellite auroral images, the drive for global knowledge has been relentless and has always delivered new and unexpected perspectives of the structures and phenomena being studied. This need for an overall perspective has brought us to the point of now being able to globally image the particle populations of the magnetosphere. This, combined with existing large-scale observations, provides for the first time, the capability of observing the magnetosphere on a global basis. We expect that when this capability is realized, our concepts of the magnetosphere will be altered every bit as dramatically as Professor Van Allen's discovery of the radiation belts changed our view of the space environment.

1. INTRODUCTION

The need for global observations is twofold:

(1) they put "things" into perspective,
and

(2) they show the summation of local effects.

However, words do not do justice to the need for and importance of global observations. Without them, our knowledge and understanding is severely restricted; restricted to extrapo-

Magnetospheric Physics, Edited by B. Hultqvist and C.-G. Fälthammar
Plenum Press, New York, 1990

lation and/or guesswork. We remain as limited as the proverbial blindmen examining an elephant - and in magnetospheric physics, from a global perspective, that is our present position. Our observations have been local and separated not only in space but also in time. Consequently while we have developed a good (and in some cases, a remarkable) understanding of local processes in specific locations, we have a nearly complete lack of understanding of how these processes and locales work together to form the whole - we have no accurate overall picture. To be sure, we have cartoons and schematic overviews that are valuable in organizing our results and in guiding future work. Also, we are beginning to see some very exciting results from computer simulations that present another valuable perspective of the magnetosphere and its dynamics. However, without actual global observations, cartoons and simulations represent our global horizons.

The need for global observations is not a new concept - I am only stating in my own way the thoughts of many researchers over the past several decades. For example, ground-based researchers have been aware of the importance of global observations for over a century and continually have emphasized the use of world-wide networks of observing stations to uncover the large-scale behavior of the parameters being observed.

What is new today is the technological capability of making global observations of a variety of magnetospheric phenomena; a capability not possible but a few years ago. We shall demonstrate in the following sections the power of global observations by showing examples of present capabilities, projected capabilities, and quantitative results. We begin by showing in the next section early magnetospheric examples of the natural drive to place local observations into a global perspective.

2. EARLY EXAMPLES

Figure 1a shows the response of the Explorer 1 University of Iowa geiger tube as it saturated when it encountered the earth's radiation belts. These data represent the discovery of the intense radiation contained in the earth's high altitude magnetic field, the Van Allen radiation belts (Van Allen et al., 1958). It is for this discovery and his pioneering work in establishing the field of magnetospheric physics that we are celebrating Prof. James A. Van Allen, the 1989 Crafoord Prize honoree, at this symposium - a celebration richly deserved and long overdue.

Following the discovery of the radiation belts, Van Allen et al. (1959) synthesized available data and presented the first extrapolation towards a global view of the earth's charged particle population. This is illustrated in Figure 1b where the actual data are shown as the solid lines and the global view obtained by extrapolating these results along magnetic field lines is given by the dashed lines. A two peak structure became apparent when high apogee altitudes were attained and the macroscale extrapolation of these later results is shown in Figure 1c (Van Allen and Frank, 1959). Figure 2 summarizes this early work and shows a global view of the radiation belts as generally accepted in the late 1950's and summarized at a 1959 AGU symposium (Newell, 1959).

We see that the need for global perspectives was in evidence right at the outset of the space age. Note that these extrapolations to a macroscopic view were based on sound physical principles describing the behavior of charged particles trapped in the earth's magnetic field. Under the fundamental assumptions used, that the earth's magnetic field in space was known and that it was static, these views were thought to be reasonably accurate. We now know that the earth's magnetic field in space is neither static nor well-known. Consequently, these early global views require modification.

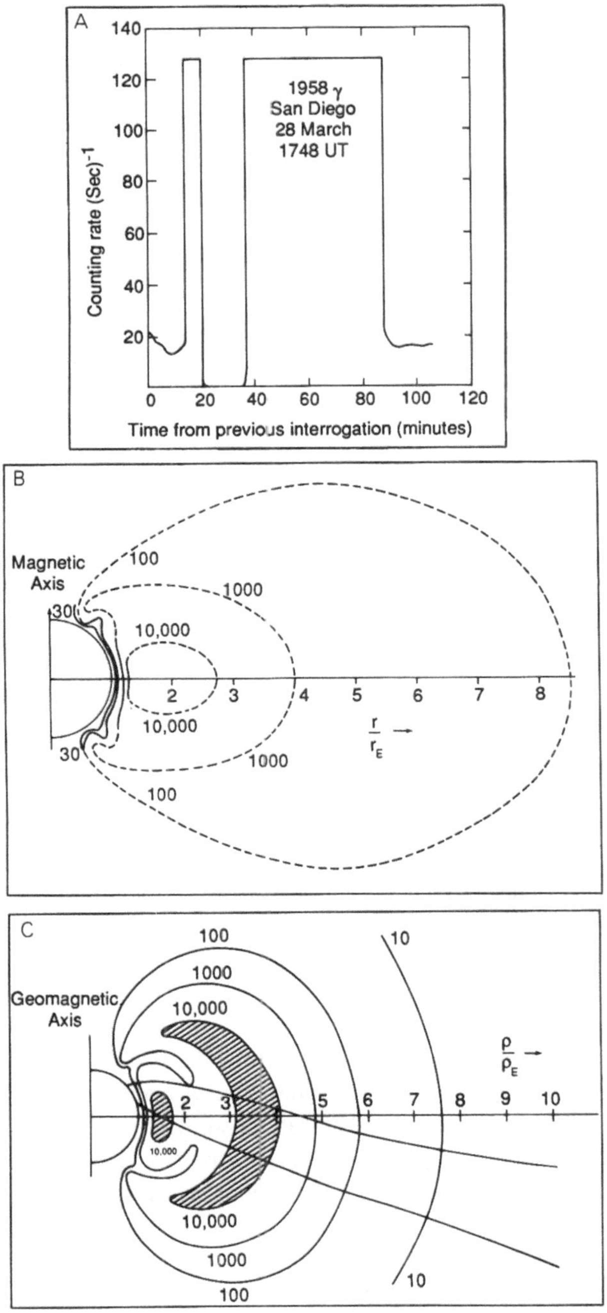

Fig. 1a). Saturation of University of Iowa Geiger tube as it encountered the high intensities of the earth's trapped radiation (from Van Allen et al., 1958).

Fig. 1b). Early global diagram of the trapped radiation. Solid lines show available data and dashed lines show extrapolation to high altitudes (from Van Allen et al., 1959).

Fig. 1c). Data from high altitudes along the orbital track shown indicated a two peak spatial structure for the radiation region. This structure was incorporated into the global picture as shown (from Van Allen and Frank, 1959).

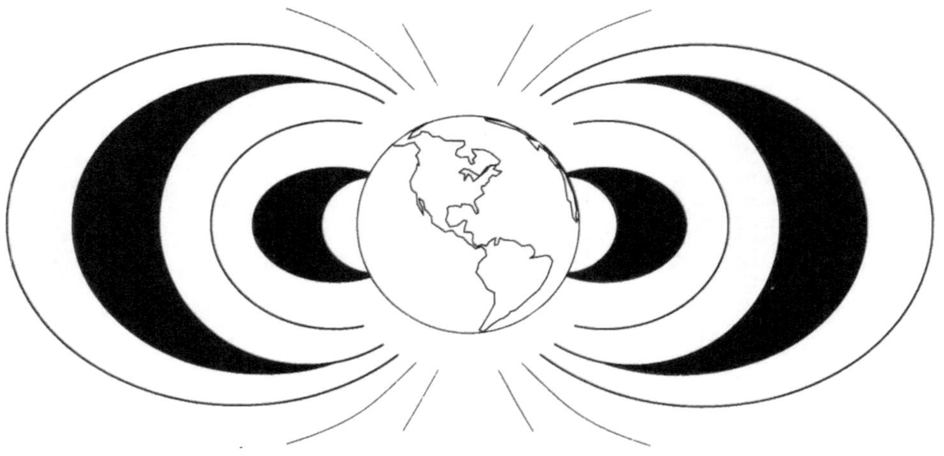

Fig. 2. The global view of the radiation belts as generally accepted in 1959.

Figure 3 shows a global cartoon of the magnetosphere that has evolved from three decades of local observations. It is a remarkable picture in that it represents in many ways a brilliant synthesis of a huge number of ground-based, rocket, and satellite observations made at widely disparate places and times. Processes, structures, and characteristic features of the earth's magnetosphere, not dreamed of thirty years ago, are summarized in this figure. Indeed this object, a magnetized plasma, is now known to be a fundamental structure of the cosmos and may represent a basic step in the cosmological evolutionary process. However, it remains a cartoon - we do not yet have direct observations of this most interesting object.

The following sections will present examples of presently available and projected macroscale and global observational capabilities. We will attempt to demonstrate both their qualitative global importance as well as their quantitative possibilities.

3. UPPER ATMOSPHERE/IONOSPHERE

Figure 4 shows a ground-based view of an aurora - for decades our only view, limited to the available sky over an observing site. From these views many important auroral characteristics were obtained. Although certain global features were inferred by utilizing simultaneous observations from many stations, an accurate global perspective of the aurora remained elusive.

Figure 5 shows the beginnings of our present global auroral perspective. This came about through the inclusion of cameras on low-altitude polar-orbiting satellites (in the case of Figure 5, the United States DMSP series) that were sensitive to auroral emissions. Figure 5 shows a collage of seven consecutive DMSP passes that together display clearly the overall spatial extent of the aurora. However, these seven passes span a time period of over ten hours, while auroral phenomena occur on a much faster time scale (≤ minutes). Thus Figure 5 represents a complex sampling of auroral regions from which it is difficult to extract accurate global behavior beyond that seen in one pass (one of the panels in the figure).

The era of true global auroral observations began with launch of the DE-1 satellite on 3 August 1981. DE-1 was launched into an initial polar orbit having a 570 km perigee altitude, a 3.65 R_E apogee altitude, an apogee latitude of 78.2° N and geographic local time of 0200 hours. It carried onboard the University of Iowa auroral imager; an instrument that was able to observe the aurora with a time resolution of 12 minutes (Frank et al., 1981). This remarkable instrument paved the way for a rich harvest of superb global auroral observations such as obtained from the Viking, EXOS-D, HiLAT, and Polar Bear satellites. Figure 6 shows a typical DE auroral image obtained at vacuum-ultraviolet (VUV) wavelengths 123-155 nm, for which the principal emissions from the aurora and atmospheric dayglow (upper right) are due to atomic oxygen at about 130.4 and 135.6 nm and to the LBH band of molecular nitrogen. The nearly instantaneously observed global character of the aurora is clearly evident. Time sequences of such images provide otherwise unavailable information on the dynamics of the entire auroral region. To demonstrate the gain in perspective such images give, the white circle represents the field-of-view of a ground-based all-sky camera at Kilpisjärvi, Finland.

It is axiomatic that important, new information always accompanies the first realization of global observational capabilities. For example, Figure 7 shows an image of the theta aurora, a configuration containing a bright linear auroral structure (a transpolar arc) bisecting the normal auroral oval. This image obtained at 0022 UT on 11 May 1983 shows one of the finest examples available to date of this type of auroral distribution. The transpolar arc extends across the polar cap from local midnight to local noon. Motion of this arc in the dawn-dusk direction appears to be controlled by the direction of the B_y component of the interplanetary magnetic field (IMF). In the Northern Hemisphere, the arc moves in the direction of the IMF B_y component.

There are many examples of excellent quantitative results that can be obtained from such images - too many to discuss within the limits of this report. For example, auroral images obtained in a number of different wavelengths, including x-rays, provide the only practical way to measure quantitatively the global energy deposition into the upper and middle atmosphere due to magnetospheric electron precipitation. Further, auroral images provide a measure of the

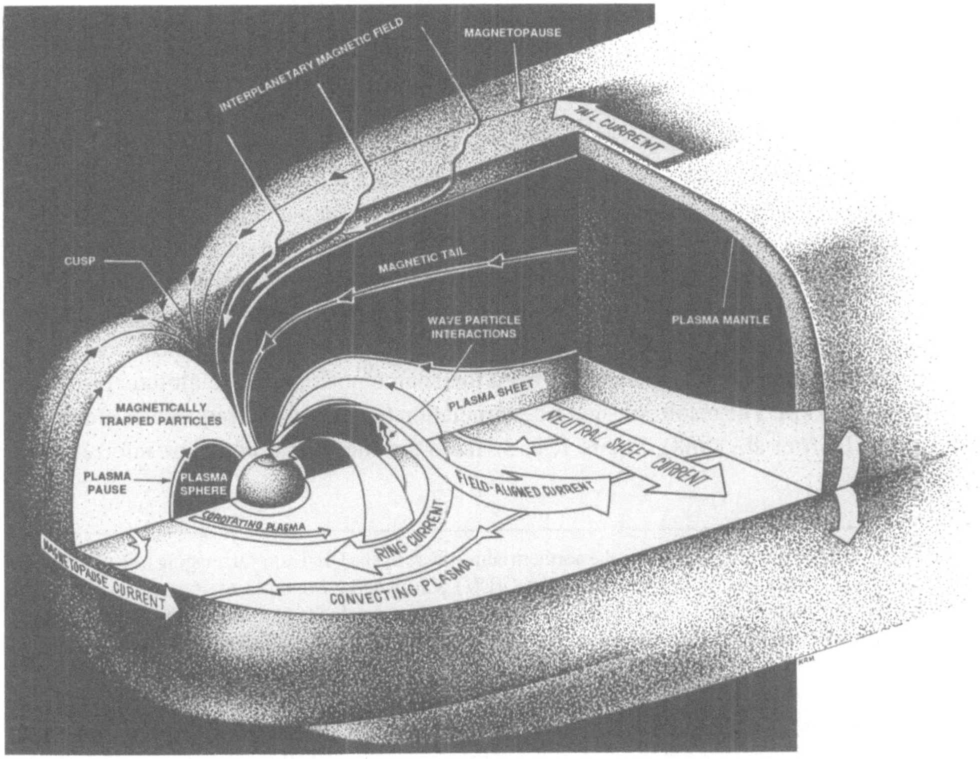

Fig. 3. Global schematic of the magnetosphere in common use today.

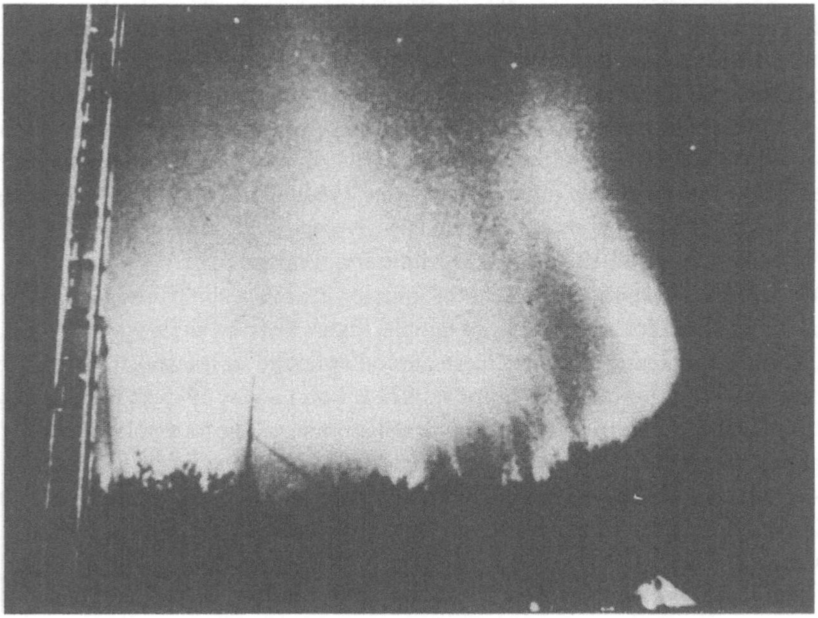

Fig. 4. A ground observer's view of an aurora.

area of the polar cap and its time variation which in turn gives a measure of the magnetic energy content and its time variability in the magnetotail; they provide the possibility of globally mapping ionospheric regions into the high altitude magnetosphere; they provide a physically meaningful global perspective and framework in which to place observations of ionospheric ion drifts and upper atmospheric neutral winds. These and many more studies utilizing auroral images are discussed in the excellent papers comprising the special issue of Reviews of Geophysics, Vol. 26, No. 2, May 1988, "The Dynamics Explorer Program: 5 Years Later".

We present here one example of such a study, the tracing of auroral kilometric radiation (AKR) to its source and to associated auroral regions (see Gurnett and Inan, 1988; Huff et al., 1988). AKR is known to be associated with aurora. If AKR is generated at the electron cyclotron frequently, ω_e a unique source position can be determined from the intersection of the ω_e surface and the line along the direction of arrival, a direction that can be determined by the DE-1 plasma wave instrument. After determining the AKR source position, its relation to the aurora can be determined by tracing the magnetic field from the source position to the altitude of optical emissions, ~200 km. The results of such a determination for four wave frequencies are shown in Figure 8 (Huff et al., 1988). Points 1, 2, 3, and 4 are the source positions inferred from

---

Fig. 6. A typical auroral image taken over the northern hemisphere by the DE-1 auroral imaging instrument. This false-color image is obtained at vacuum-ultraviolet (VUV) wavelengths from 123 to 155 nm for which the principle emissions from the aurora and upper-atmosphere dayglow are due to atomic oxygen at 130.4 nm and 135.6 nm and to the LBH band of molecular nitrogen. For orientation, the white circle shows the coverage provided by an all-sky camera at Kilpisjärvi, Finland.

Fig. 13. Sequence of IRIS images beginning at 0710:10 hours on 15 Jan. 1988. Each panel is a 10-second sample (image) of the entire 49-beam array shown in Figure 12. Absorption is shown color-coded by color bar on the right. Image orientation is shown in the upper left and the start time for each row of 10-second images is given on the left. (Courtesy of T. Rosenburg).

Fig. 6

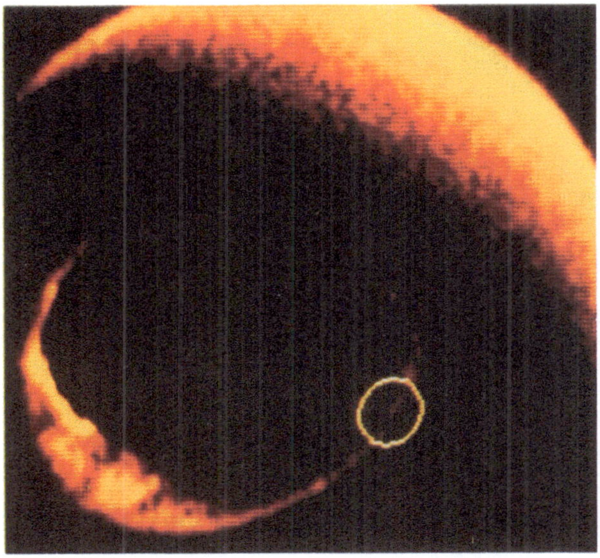

Fig. 13 South Pole IRIS, 15 January, 1988

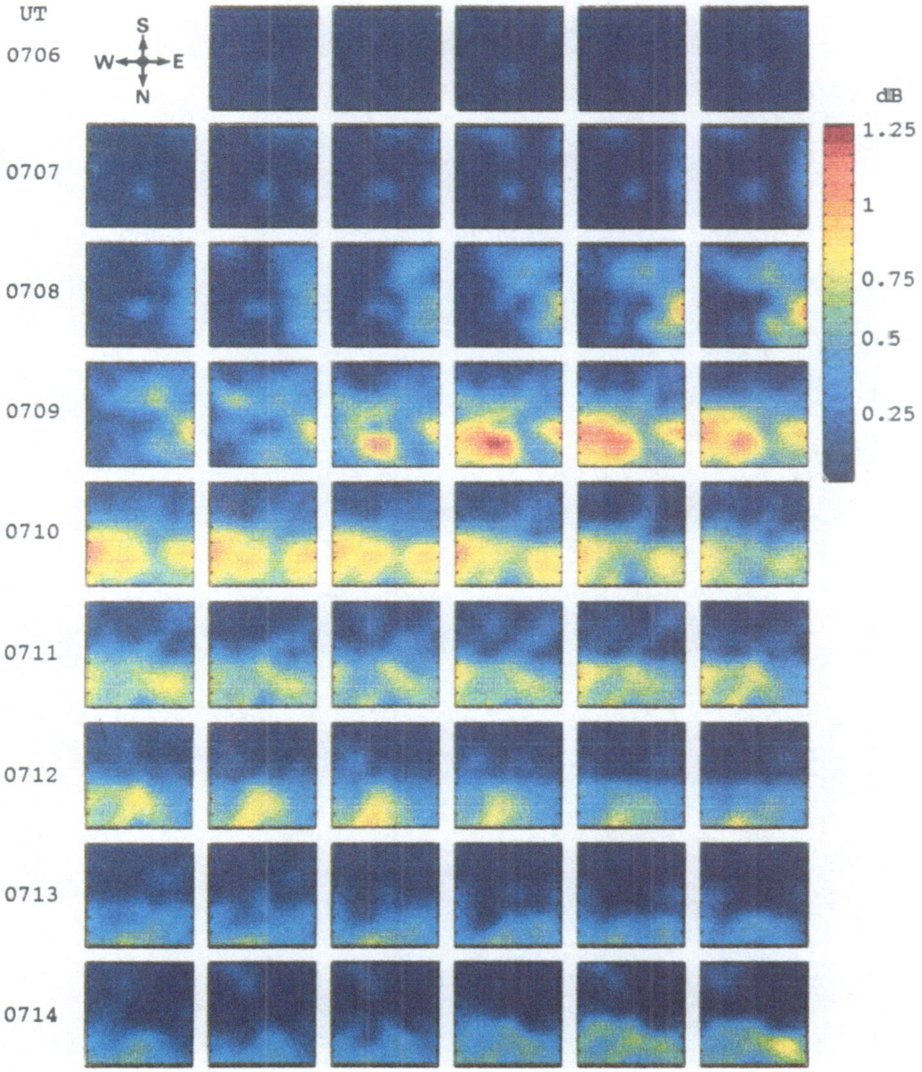

Fig. 7

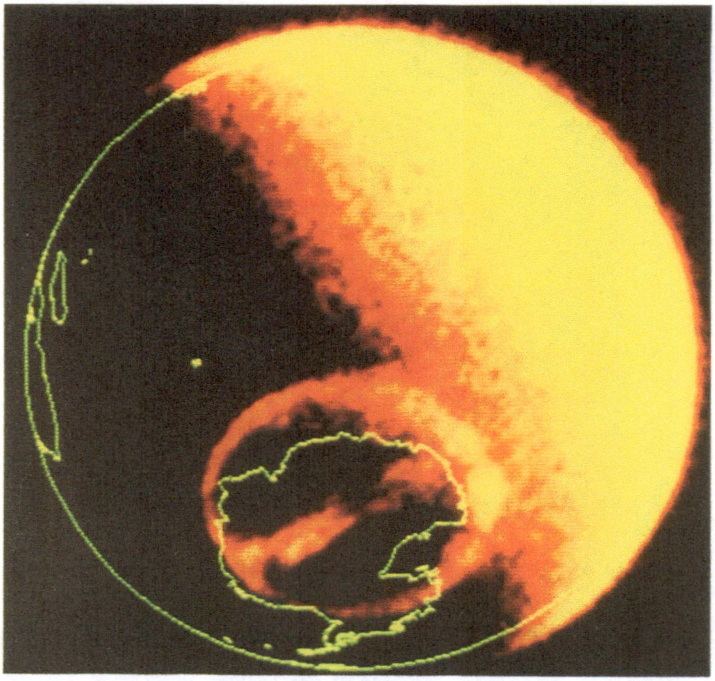

Fig. 15

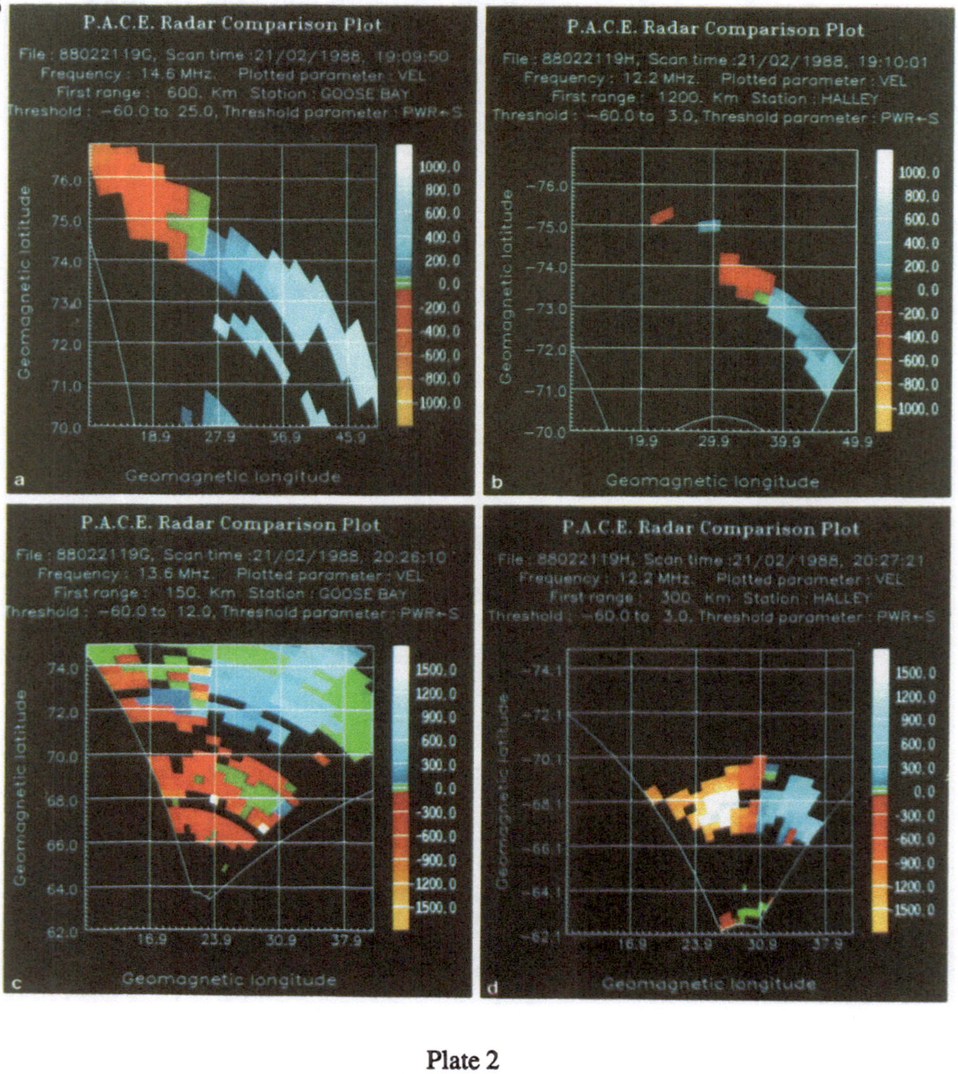
P.A.C.E. Radar Comparison Plot
File : 88022119G, Scan time :21/02/1988, 19:09:50
Frequency : 14.6 MHz. Plotted parameter : VEL
First range : 600. Km Station : GOOSE BAY
Threshold : -60.0 to 25.0, Threshold parameter : PWR+S
Geomagnetic latitude
Geomagnetic longitude
a
P.A.C.E. Radar Comparison Plot
File : 88022119H, Scan time :21/02/1988, 19:10:01
Frequency : 12.2 MHz. Plotted parameter : VEL
First range : 1200. Km Station : HALLEY
Threshold : -60.0 to 3.0, Threshold parameter : PWR+S
Geomagnetic latitude
Geomagnetic longitude
b
P.A.C.E. Radar Comparison Plot
File : 88022119G, Scan time :21/02/1988, 20:26:10
Frequency : 13.6 MHz. Plotted parameter : VEL
First range : 150. Km Station : GOOSE BAY
Threshold : -60.0 to 12.0, Threshold parameter : PWR+S
Geomagnetic latitude
Geomagnetic longitude
c
P.A.C.E. Radar Comparison Plot
File : 88022119H, Scan time :21/02/1988, 20:27:21
Frequency : 12.2 MHz. Plotted parameter : VEL
First range : 300. Km Station : HALLEY
Threshold : -60.0 to 3.0, Threshold parameter : PWR+S
Geomagnetic latitude
Geomagnetic longitude
d

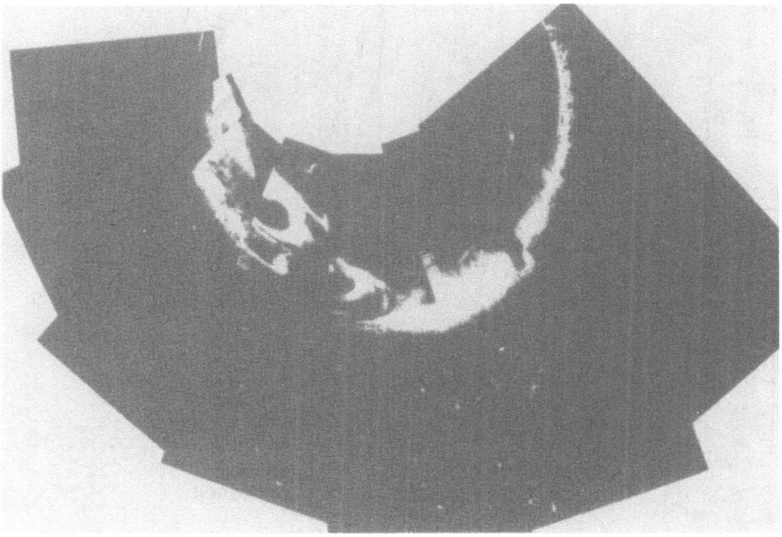

Fig. 5. The beginnings of a global observational perspective of the aurora. Auroral forms are seen in seven consecutive passes of a United States DMSP satellite and show the overall extent of the aurora, although severe time discontinuities clearly are evident.

direction-finding measurements at frequencies of 104, 136, 170, and 218 kHz respectively. The distance of the points from the Earth is related to the location of the respective cyclotron frequency surface. Figure 8 shows that the source points map along magnetic field lines to the same simultaneously observed auroral bright spot.

We mentioned earlier that ground-based observers long have recognized the importance of developing a global view of magnetospheric phenomena. It was through this recognition that a core magnetospheric behavioral characteristic, the substorm, was uncovered some 2 1/2 decades ago (Akasofu, 1965). Figure 9 shows some of the original sketches made to illustrate the progression in time of this large-scale magnetospheric phenomenon. This picture was inferred through the painstaking analysis and insightful synthesis of many simultaneous ground-station observations of the aurora and its break-up. Such efforts were the only way that global effects could be uncovered.

Fig. 7. Auroral imaging with Dynamics Explorer 1 has identified an unique auroral configuration, the theta aurora, that can occur when the interplanetary magnetic field is directed northward. This image obtained at 0022 UT on 11 May 1983 shows one of the finest examples available to date of this spatial distribution comprising the auroral oval and a transpolar arc. Motion of the transpolar arc in the dawn-dusk direction appears to be controlled by the direction of the interplanetary magnetic field along the dawn-dusk direction. In the Northern Hemisphere, the arc moves in the direction of the IMF B_y component.

Fig. 15. PACE line-of-sight velocities measured by HF (8-20 MHz) radars located in Goose Bay, Labrador (panels a and c) and Halley Bay, Antarctica (panels b and d). Velocity towards (+) and away (-) from the radar is coded in meters/second according to the color bar at the right and displayed on a grid showing geomagnetic latitude and geomagnetic longitude. The observations at ~1910 UT (panels a,b) show a velocity pattern typical of westward convection and a scattering region at the same magnetic location in both hemispheres. At ~2026 UT, two scattering regions appear and are displaced in the southern hemisphere ~5° equatorward from the corresponding northern hemisphere locations (courtesy, R.A. Greenwald).

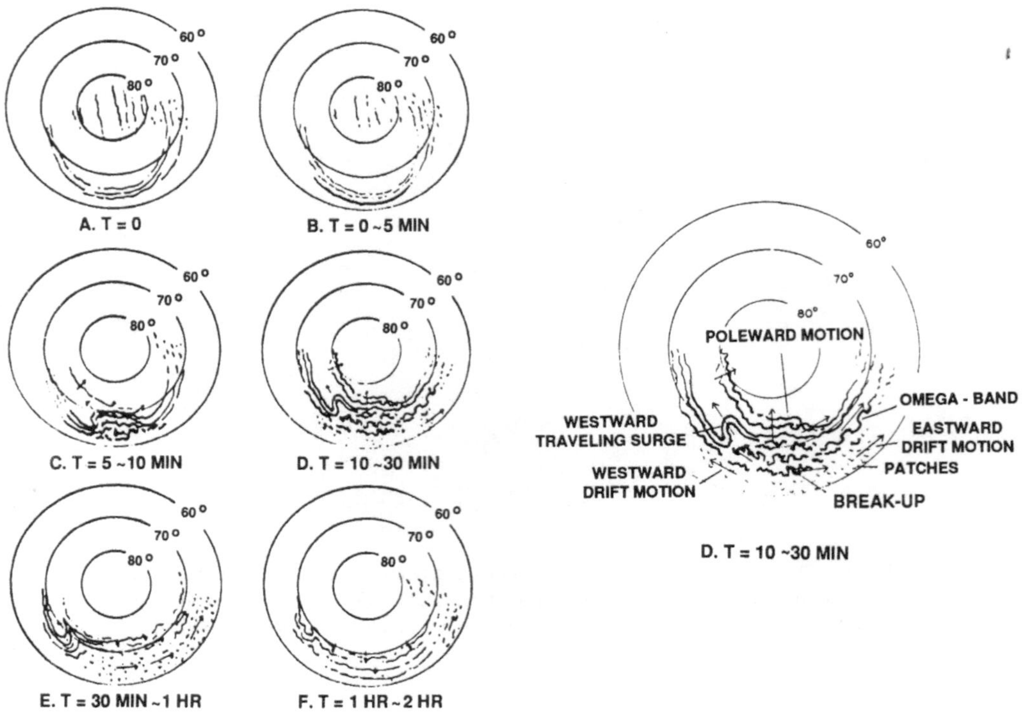

Fig. 9. Schematic diagram illustrating the development of the auroral substorm. The view is down upon the north geomagnetic pole and the sun is towards the top of the diagram. An approximate time from substorm start (T=0) is shown. An expanded diagram of substorm maximum stage is shown on the right (from Akasofu, 1965).

To compare these results with auroral images, we show a sequence of DE images during a substorm event in Figure 10. This sequence records the onset and evolution of a substorm as the DE 1 satellite descends from apogee during the period 0529 through 0755 UT on 2 April 1982. The poleward bulge of the surge is first seen in the fourth image, which begins at 0605 UT. A westward traveling surge then propagates along the poleward edge of the oval as highly structured, eastward-moving auroral forms develop in the post-midnight sector. The spacecraft location for the first image (the upper left-hand corner) is 23° N geographic latitude, 22 hours local time, and an altitude of 3.67 R_E. The spacecraft is directly over the auroral oval as the last image frame is telemetered, at an altitude of 2.17 R_E. Figure 10 shows how well the difficult synthesis of many ground-based observations (Akasofu, 1965) succeeded in unraveling the general global features of the auroral substorm. However, it is also readily apparent that these

---

Fig. 8. Comparison of AKR source position determinations with simultaneously observed auroral image. The source-surface positions determined for the four frequencies shown are all associated with the same bright auroral spot (from Huff et al., 1988).

Fig. 10. Twelve consecutive false-color images of the northern auroral oval record the onset and evolution of a substorm as the spacecraft Dynamics Explorer 1 descends from apogee position during the period 0529 through 0755 UT on 2 April 1982. Each image is telemetered in a twelve-minute interval. The poleward bulge of the surge is first seen in the fourth image, which begins at 0605 UT. A westward traveling surge then propagates along the poleward edge of the oval as highly structured, eastward-moving auroral forms develop in the post-midnight sector. General agreement with substorm development as inferred by ground-based observations and shown in Figure 9 is evident. Also readily apparent is the great improvement in our global perspective occasioned by observations such as this, particularly in the continuity of observations in both space and time.

Fig. 8

DE-1 JANUARY 27, DAY 27, 1982 0445 UT

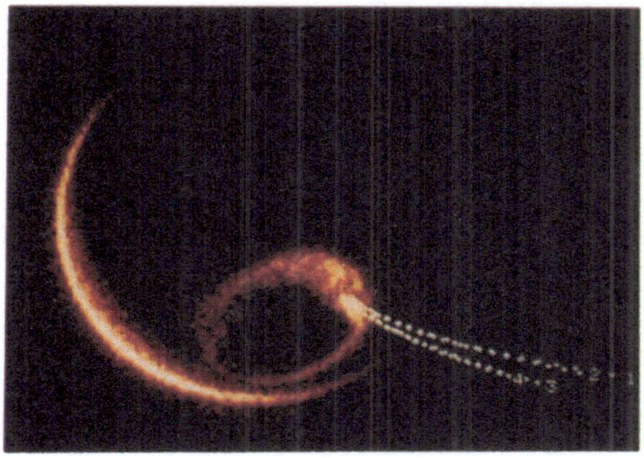

PWI

AKR SOURCE #	UT	WAVE FREQUENCY
1	C445	104 kHz
2	C445	136 kHz
3	C445	170 kHz
4	C445	218 kHz

Fig. 10

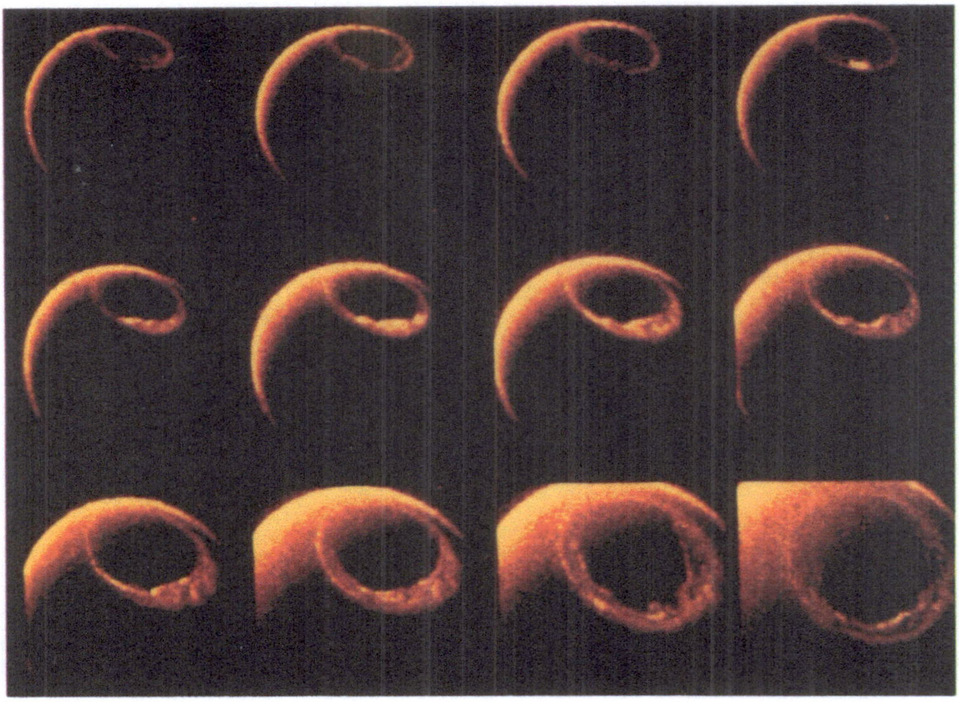

Plate 3

relatively instantaneous global auroral images greatly enhance and expand the ground-based network—particularly in spatial coverage and in the continuity of the observations both in space and time.

Another area in which ground-based observational networks have been used to provide a global view is in studies of the earth's ring current—that current responsible for ground level depressions of the earth's magnetic field occurring during events known as magnetic storms. Figure 11 shows ground-based observations obtained during the great magnetic storm of 13 September 1957. The figure shows D_{st} (a measure of the magnetic field depression, in nT) as a function of time for several low latitude stations. A global view also is shown in terms of contour plots obtained from the station network for three specific times during the magnetic storm. How these global responses of the Earth's surface magnetic field relate to the magnetospheric ring current is not known because as yet there are no global images routinely available for this fundamental current system.

Next, we consider examples of recent advances in obtaining macroscale observations of ionospheric phenomena from ground-based facilities. There has been a striking increase in the capability of observing both the spatial and temporal development of electron precipitation regions by ground-based riometers. Most measurements of cosmic radio noise absorption in the polar regions have been made with riometers using broadbeam antennas. Only limited information about the spatial structure and dynamics of energetic electron precipitation regions can be obtained by this means. Recent trends in riometry have been toward the use of multiple, narrow beam antennas operated in a fixed-beam or one-dimensional scanning mode to examine smaller ionospheric regions. A further step in this direction has been taken with the development of a phased-array imaging riometer system by T.J. Rosenberg at the University of Maryland. The first Imaging Riometer for Ionospheric Studies, IRIS, is now installed and operating continuously at South Pole Station, Antarctica. IRIS operates at a frequency of 38.2 MHz and produces 49 independent narrow beams (12 degrees full width) arranged in a 7 by 7 square array centered on the zenith, with beam peaks out to about 45 degrees zenith angle along the principal axes. The ionospheric sampling grid is illustrated in Figure 12 along with the area sampled by a standard broadbeam riometer. The two-dimensional viewing region of IRIS at D-region heights covers an area approximately 200 km on a side, with a resolution on the order of 20 km. The entire image is scanned once a second.

Figure 13 displays a succession of "images" obtained by IRIS in which the spatial and temporal development of an electron precipitation event is readily discernible. Each panel represents an image formed by the grid pattern shown in Figure 12 in which absorption levels are color-coded according to the color bar on the right. Each image is taken over a 10 second interval and the start of each minute is indicated on the left. The top left panel has been used to show the orientation of the images. Analysis of the data (T. Rosenberg, personal communication) shows a westward propagating feature traveling at a speed of ~1 km/sec. Instrumentation like IRIS will prove to be very valuable in separating spatial and temporal features of electron precipitation events over large ionospheric areas.

Another development in observing large-scale ionospheric features has come from the Polar Anglo-American Conjugate Experiment (PACE) conducted by J.R. Dudney of the British Antarctic Survey and R.A. Greenwald of The Johns Hopkins University (Baker et al., 1989). PACE uses coherent scatter hf radars operating at 8-20 MHz and located at Goose Bay, Labrador and Halley Bay, Antarctica to observe ionospheric irregularities simultaneously over large conjugate areas. The field of view of each radar system extends in absolute value from approximately 64° to over 85° magnetic invariant latitude. The overlap of the conjugate fields-of-view is shown in Figure 14. The total conjugate area is in excess of $3 \cdot 10^6$ km^2 and includes other important research facilities such as those at South Pole and Søndre Strømfjord.

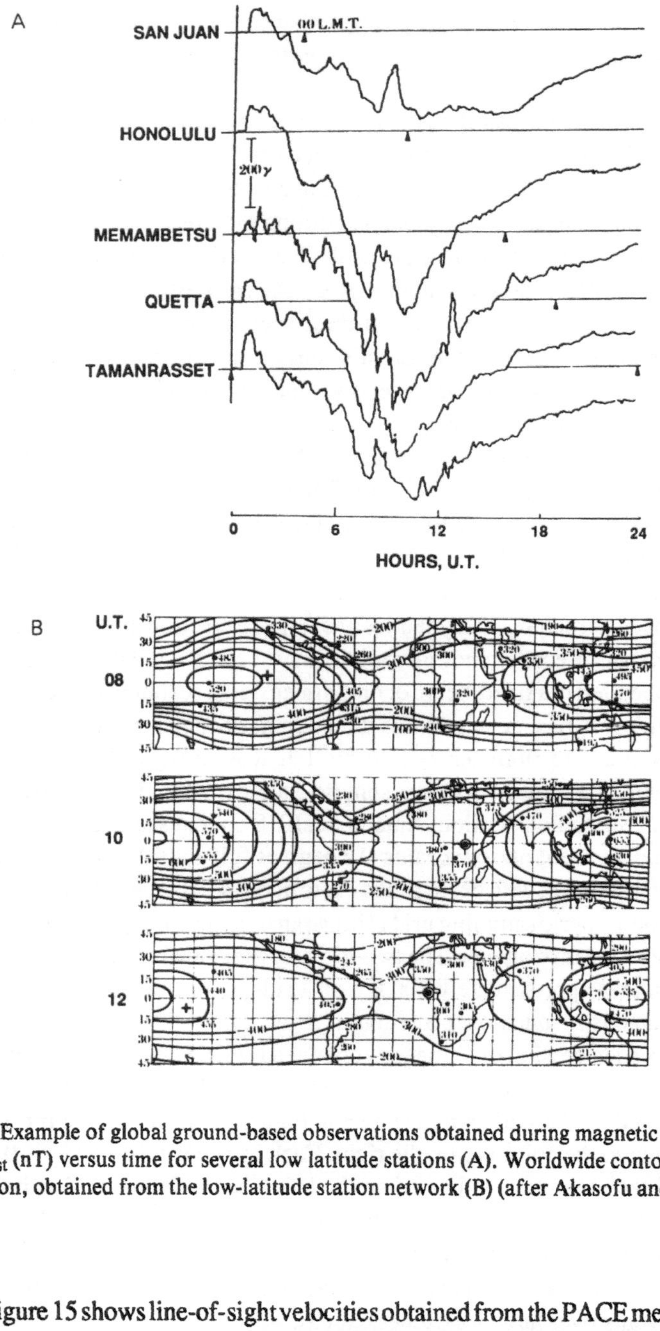

Fig. 11. Example of global ground-based observations obtained during magnetic storm of 13 September 1957. D_{st} (nT) versus time for several low latitude stations (A). Worldwide contours of the surface field depression, obtained from the low-latitude station network (B) (after Akasofu and Chapman, 1972).

Figure 15 shows line-of-sight velocities obtained from the PACE measurements at ~1910 UT, 21 February 1988 (Dudney and Greenwald, personal communication). The light-of-sight velocity is coded in magnitude (m/s) and direction (+, towards; -, away) according to the color bar at the right of each panel and is displayed on a geomagnetic latitude-longitude grid. Panels a) and c) show the Goose Bay results and panels b) and d) show the Halley Bay results. The magnetic local time for these observations is ~2200 hours MLT.

Panels a) and b) show velocity patterns at both radars that are typical of westward convection and a scattering region that is located at approximately the same geomagnetic location in each hemisphere. Panels c) and d) show PACE results obtained ~1 1/2 hours later at which time both radars observe two distinct regions; a poleward region characterized by high speed westward convection, and an equatorward region characterized by low negative Doppler velocities. In this case, the southern hemisphere scattering regions seem displaced equatorward by ~5° relative to their northern hemisphere counterparts. Such results will provide stringent tests for global models of the magnetospheric electric and magnetic field configuration.

4. MAGNETOSPHERE

The preceding discussion described examples of global and/or macroscale geophysical observations of the ionosphere, upper atmosphere, and the Earth's surface magnetic field. In this section, we describe briefly global observations of high altitude magnetospheric regions that are expected to be available in the future.

Advances in optical instrumentation, particularly in threshold sensitivities, make it feasible to remotely sense tenuous plasmas by detecting their optical emissions (Broadfoot, 1986). Such measurements provide the capability of globally observing low energy ion populations in the magnetosphere. For example, the use of intensified charged coupled devices with spectrographic and photometric instrumentation provides sensitivities sufficient to image the plasmasphere by measuring resonantly scattered He^+ emissions at 30.4 nm. From an appropriately positioned satellite platform, it then becomes possible to observe the global morphology and dynamics of the plasmasphere. Aside from H^+, He^+ is the most abundant plasmaspheric ion and thus is a natural candidate for remotely imaging the plasmasphere. To illustrate the capability of a plasmaspheric image utilizing the He^+ 30.4 nm emission line, we show in Figure 16, expected emission maps based on the two He^+ models shown. Model 1 stems

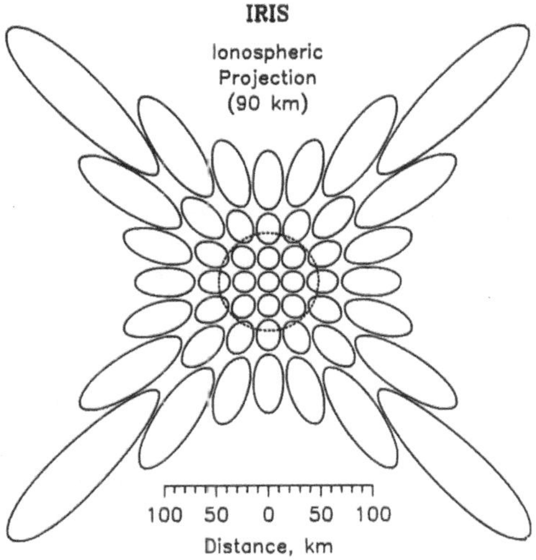

Fig. 12. Projection to ionospheric heights of the beam patterns of the 49 beam array of IRIS (Imaging Riometer for Ionospheric Studies). This array, installed at South Pole Station during the 1987-1988 austral summer season, employs a 12° full width for each beam, operates at a frequency of 38.2 MHz and is sampled (all 49 beams) in 1 second. The dotted circle in the center of the array shows for comparison the pattern of a single wide beam riometer. (Courtesy of T. Rosenburg).

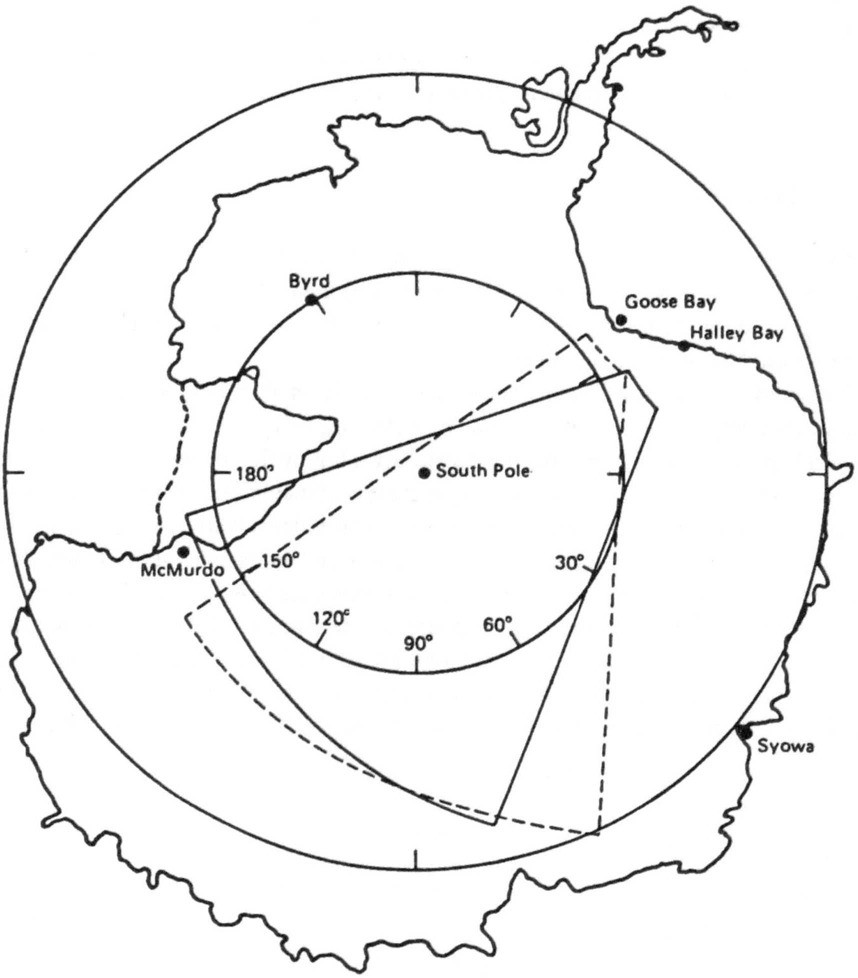

Fig. 14. Conjugate overlapping areas of the Polar Anglo-American Conjugate Experiment (PACE) radars projected over the Antarctic continent. Total conjugate area is $3 \cdot 10^6$ km² and includes the South Pole viewing area of IRIS shown in Figure 12 (from Baker et al., 1989).

from DE-1 measurements obtained during magnetically disturbed conditions (Horwitz et al., 1984) and model 2 is an equilibrium model based on ISEE measurements (Waite et al., 1983). The peak 30.4 nm emission intensity differs by a factor of ten for the two different models. The application of standard deconvolution techniques to these line-of-sight images can be used to infer details of these (or any existing) He^+ distributions as a function of L.

Figure 17 shows views of the plasmasphere based on Model 1 as seen from observing positions at the noon meridian both on and 35° above the magnetic equator. The brightest He^+ 30.4 nm emission occurs near the Earth where the He^+ column density through the plasmasphere reaches its maximum value. Measurable emission should be seen out to ~5 R_E based on this model. The image from 35° magnetic latitude shows a view of the full longitudinal extent of this model plasmasphere.

Imaging the entire plasmasphere as illustrated in Figures 16 and 17 is an excellent way to monitor global changes in plasmaspheric ion density and spatial distribution. Since full plasmasphere images can be obtained on time scales of minutes, variations in response to geomagnetic activity can be tracked readily. For example, the refilling of newly-corotating flux tubes outside of the previously existing plasmasphere can be studied by such imaging. Global images during active periods will show directly how much the plasmasphere is compressed and how much of it is detached and drifts through the magnetosphere under the influence of the ambient electric and magnetic fields. Plasmaspheric images also may allow a mapping of the inner edge of the magnetospheric convection process. These examples illustrate the value of plasmaspheric global images as a major new capability in future magnetospheric research.

Another resonantly scattered photon is the 83.4 nm scattered radiation from O^+, again giving the possibility of observing the distribution of low energy plasma in the magnetosphere (Swift et al., 1989). The 83.4 nm radiation is an attractive choice since O^+ is generally the second most abundant ion in the magnetosphere. Figure 18 shows a simulation of what an image in 83.4 nm might show using a simple model of O^+ plasmas in an asymmetric magnetosphere. In this model, the plasmasheet is represented as a slab. The view shown is from lunar orbit at 1400 hours local time and 30° above the ecliptic plane. An instrument with a 40°x40° field-of-view and a resolution of 0.4° was assumed for this simulation. Since this resolution is considerably better than that attainable for instruments in the foreseeable future, Figure 18 actually shows an idealized version of an 83.4 nm image. Also, ionospheric O^+ ions were blanked out below 1.5 RE in order to avoid the extremely bright core they would contribute to the image. While this is a promising technique, more study is needed to test thoroughly its feasibility.

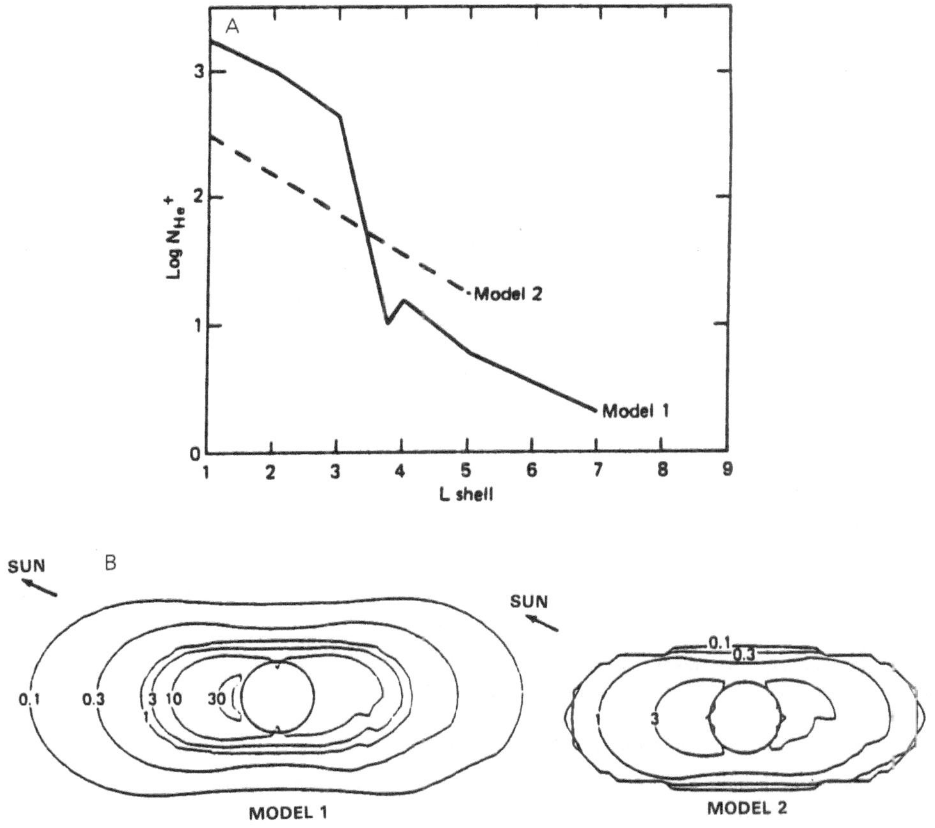

Fig. 16. Top: Model plasmaspheric He^+ densities used to estimate 30.4 nm intensities. Model 1 (Horowitz et al., 1984) represents magnetically disturbed conditions and Model 2 (Waite et al., 1983) is an equilibrium model.
Bottom: Plasmaspheric He^+ 30.4 nm emission maps for the two models shown above. Comparison shows that present instrumentation can readily track the plasmasphere's evolution from quiet to disturbed conditions.

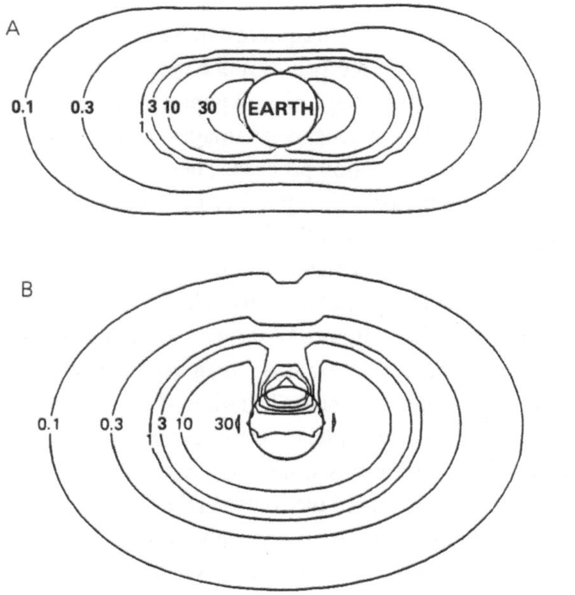

Fig. 17. Plasmaspheric He^+ 30.4 nm emission maps as observed from indicated positions. (a) Isophotes for distribution Model 1. Seen from a point on the Earth-Sun line in plane of the magnetic equator. (b) The same as in (a) but seen from a point in the noon meridian at a magnetic latitude of 35°.

Finally, we discuss another global-magnetosphere observational technique, one whose feasibility recently has been demonstrated, namely the measurement of energetic neutral atoms (ENA) to provide global images of the magnetospheric energetic particle (hot plasma) populations. The basis of this technique rests on the fact that the charged particle populations of the magnetosphere are embedded in a tenuous atmosphere of cold, neutral hydrogen atoms, the earth's hydrogen geocorona. ENA are produced by the charge exchange of the energetic ions in magnetospheric plasmas with these hydrogen atoms. The ENA travel along straight line paths from the point of charge exchange and can be detected to form an image of the source magnetospheric plasmas in a manner analogous to the imaging of aurora by the detection of photons emitted from the upper atmosphere.

The concept of measuring ENA and using their detection to image energetic ion populations in the magnetosphere has been tested successfully with measurements from instruments designed primarily for energetic charged particle observations on the IMP and ISEE spacecraft (Roelof et al., 1985; Roelof, 1987). Roelof et al. (1985) presented the first

Fig. 18. Simulation results showing resonantly scattered 83.4 nm radiation from magnetospheric O^+. A simple model of magnetospheric O^+ distribution including a slab approximation for the plasmasheet was used. In addition O^+ below 1.5 R_E was not considered due to the intense glow that would be introduced. The view is a 40°x40° view from lunar orbit at 30° above the ecliptic plane (from Swift et al., 1989).

Fig. 19. Models, simulation, and measurement of an ENA image of the storm-time ring-current ion population, represented in the earthward hemisphere viewed from the ISEE-1 spacecraft. Closed contours show equatorial crossing locus of L=3 and 5 field lines. These contours all connected every 3 hours in magnetic local time with the L=3 and 5 field lines drawn in at noon (to the right), dusk, midnight (to the left), and dawn. (a) Column-integrated ion differential flux from the model ring current; (b) ENA flux computed from the model ring-current ion distribution in panel a; (c) Simulated MEPI rates (average counts per pixel) computed from model ENA fluxes of panel b; (d) Actual MEPI image from 0955 to 1000 UT, 29 September 1978. MAX values are color-bar maxima (from Roelof and Williams, 1988).

Fig. 18

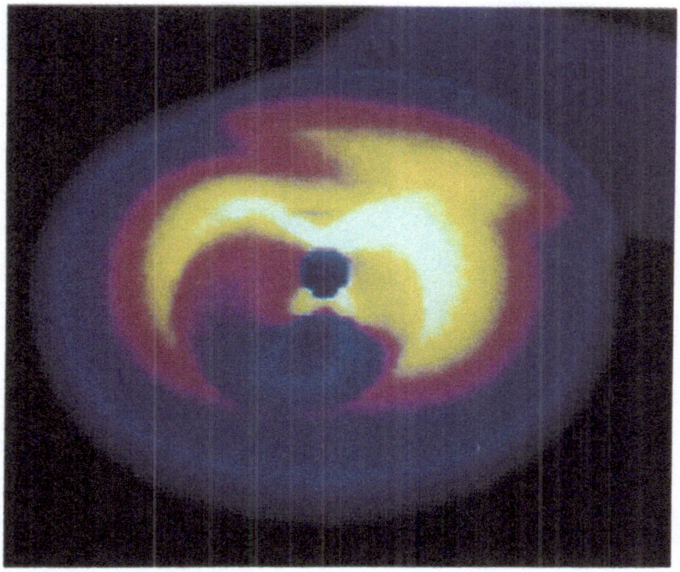

Fig. 19

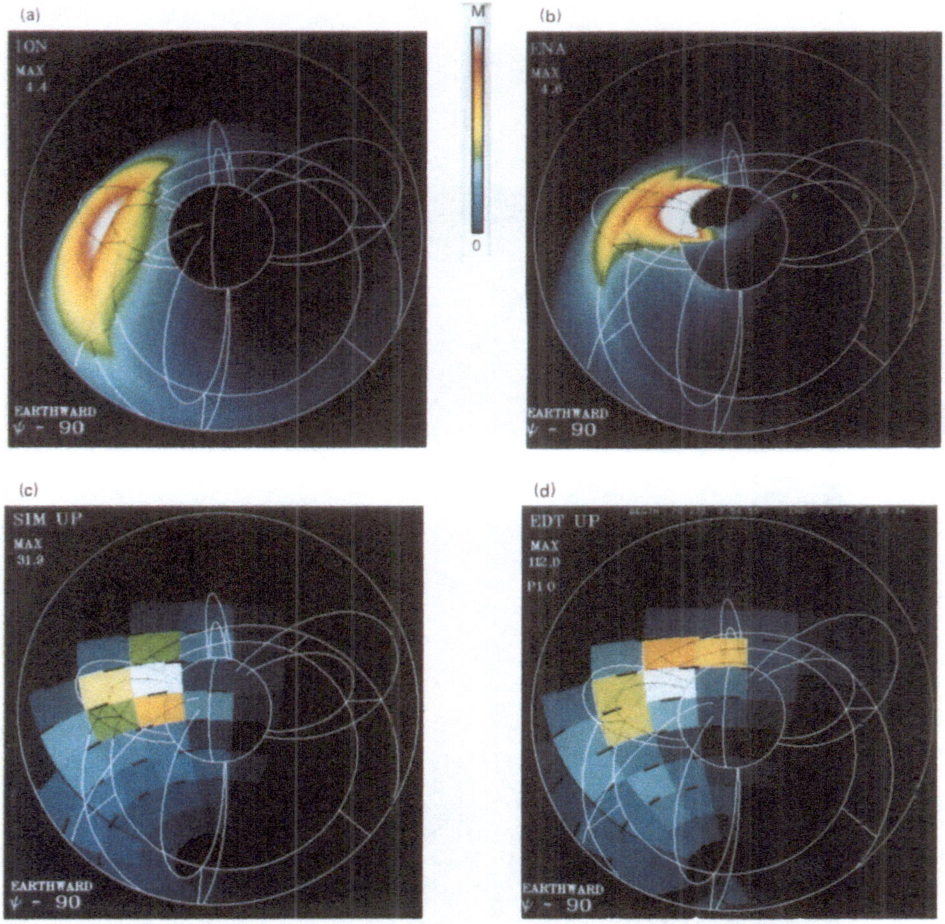

Plate 4

Fig. 20

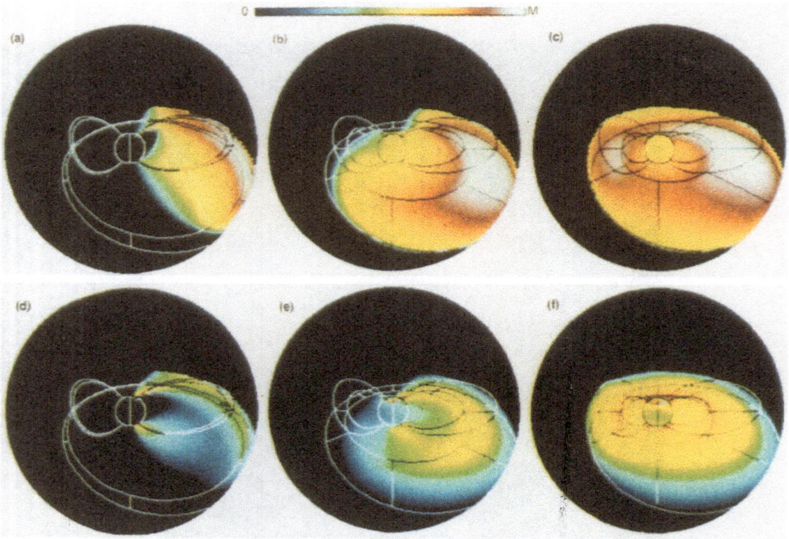
0
M
(a)
(b)
(c)
(d)
(e)
(f)

Fig. 21

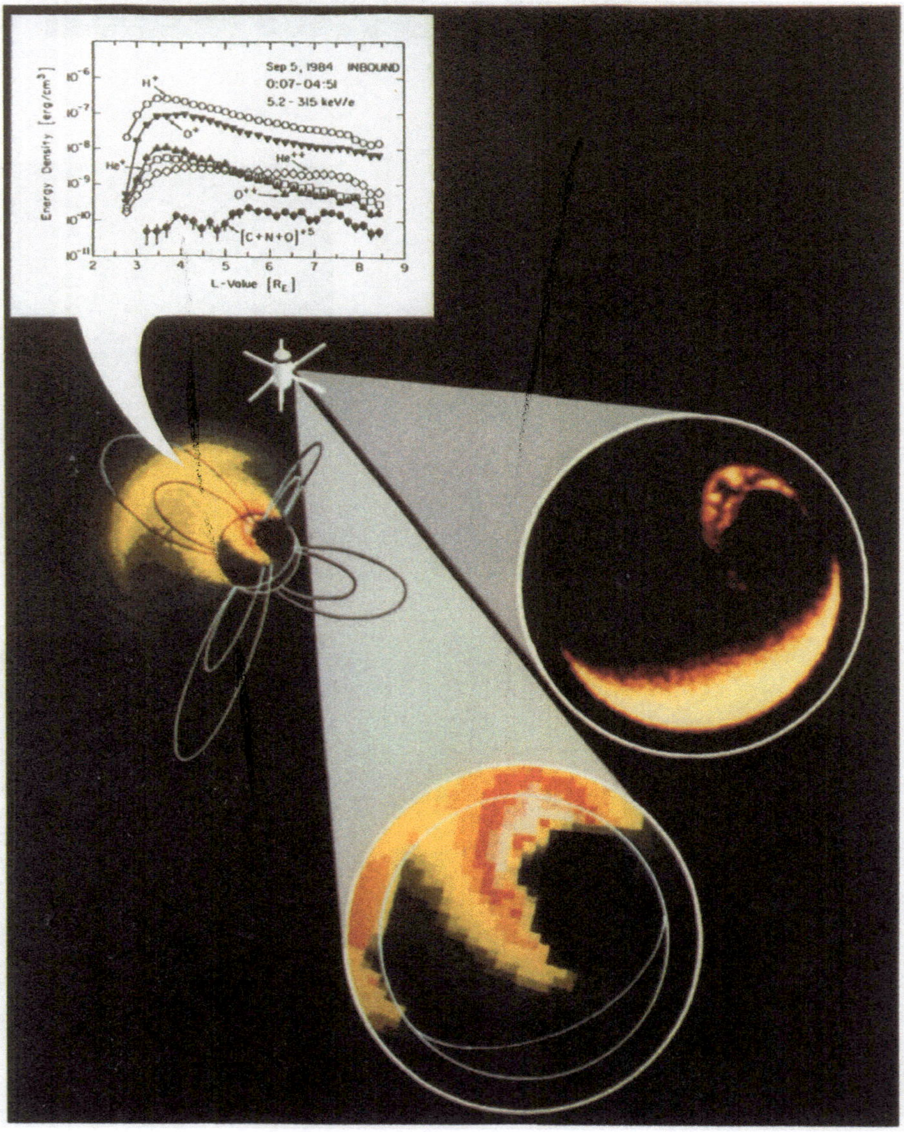
Sep 5, 1984 INBOUND
0:07-04:51
5.2 - 315 keV/e
H+
O+
He+
He++
O++
[C+N+O]+5
Energy Density [erg/cm³]
L-Value [R_E]

satellite-based observations of the decay of the global ring current by means of detecting ENA (most probably oxygen atoms) generated via charge exchange. Roelof (1987) extended these observations and obtained the first-ever global image of the ring current by iterating ring current distributions in a model magnetic field until their convolution through charge exchange with the hydrogen geocorona produced the ENA patterns observed. Further details describing the production of ENA in the magnetospheric energetic ion distributions are given in Roelof and Williams (1988), McEntire and Mitchell (1988), and Keath et al. (1988).

Figure 19 summarizes the analysis of Roelof (1987) yielding the first global image of the magnetospheric ring current. The data (bottom right panel) are from the ISEE 1 Medium Energy Particles Instrument (MEPI) which measured the three-dimensional angular distributions and energy spectra of ions and electrons of energies greater than approximately 22 keV. The data in Figure 19d pertain to the main phase of a magnetic storm on 29 September 1978 when the surface magnetic field depression was -241 nT. At that time, ISEE 1 was at a magnetic latitude of 51.5° N, at 0730 hours magnetic local time, and a radial distance of 2.6 R_E.

All panels in Figure 19 are "fish-eye" projections of the earthward hemisphere of the sky. The Earth disk and its terminator are outlined; circles representing the magnetic equator at 3 and 5 R_E are connected by radial lines every 3 hours of magnetic local time, including noon (which is to the right). Magnetic field lines for L = 3 and 5 are drawn in the planes of magnetic noon, dusk, midnight (to the left), and dawn (toward the reader). A common linear color bar is used, and its maximum value is indicated in each panel.

The ENA data are given in Figure 19d. The MEPI on ISEE 1 scans the entire sky in 12 spins of the spacecraft. Data are grouped into 16 angular sectors of 22.5° during each spin, so the ENA flux is binned into 192 "pixels", each 22.5° by 13.3°. The MEPI energy channel used corresponds to 24 to 34 keV for H and 60 to 77 keV for O. The ENA image in Figure 19d reveals the very dramatic midnight-noon asymmetry, and thus this single image immediately established that the main-phase, storm-time ring current was strongly asymmetric.

The extraction of the actual ion fluxes and their spatial distribution from the observed ENA image (Figure 19d) requires a computer simulation using a model ring current ion distribution. The model ring current distribution, shown in Figure 19a as the line integrated flux to the ISEE 1 position, is convolved with the geocoronal hydrogen distribution (Rairden et al., 1986) to predict the ENA image intensities expected at ISEE 1. These ENA intensities are shown in Figure 19b. This model ENA flux is then passed through the MEPI telescope function so that a direct pixel for pixel comparison can be made between the predicted (Figure 19c) and measured (Figure 19d) ENA flux. This process is iterated until a best match to the data is

Fig. 20, (a to c). False-color images of model ring current ion intensity evolution. Color represents the column-integrated flux of ions whose instantaneous velocity vector points toward the viewing point. Panels correspond to different times after ion injection. Closed contours indicate the equatorial crossing locus of the inner and outer L shells containing the model fluxes. Magnetic local time cuts at every 3 hours (45°) connect the equatorial loci, with 1800 MLT being the lower vertical line; the sun is to the left. The linear color bar runs from zero to the maximum (M). The viewing point is at a radius of 8 R_E and 30° geomagnetic latitude in the dusk meridian plane. (d to f). 0False-color images of the ENA flux arising from the modeled interaction of the ions in (a) to (c) with the hydrogen geocorona. Same format as for (a) to (c) (from Roelof and Williams, 1988).

Fig. 21. Global imaging and in-situ observations. A DE-1 auroral image is shown along with a simulated ENA image based on ISEE-1 results obtained during the main phase of the 29 September 1978 magnetic storm. The ENA images shown have been simulated from the DE-1 position. Both an ENA low-altitude close-up view and an overall magnetospheric view are shown in comparison with the auroral image. In-situ charge state observations of the Earth's ring current (Gloeckler et al., 1985) are shown to exemplify the unique capability of magnetospheric physics in having both global and simultaneous local observations available in the future.

obtained. The inferred ring current distribution shown in Figure 19a yields an agreement with the observed ENA fluxes of better than ~10% for each MEPI pixel. The model ring-current ion distribution finally obtained (Figure 19a) from the ISEE 1 image of 0955-1000 UT, 29 September 1978, (Figure 19d) has the ions confined to field lines of $3 < L < 5$ in the nightside hemisphere, consistent with general expectations for the main phase of a geomagnetic storm.

We have discussed these initial ENA imaging results (Roelof, 1987) in detail in order to underscore the fact that they show clearly the feasibility of using ENA to obtain global images of the magnetospheric energetic particle populations. In fact, present ENA camera designs provide sensitivities sufficient to image the quiet-time ring current, the inner regions of the plasmasheet, and substorm injections with time resolutions to ~1 minute. These images also will allow a measure of the distorted geomagnetic field configuration and a sequence of images will allow a measure of the global electric field. Used with the global observations discussed earlier, ENA images will allow an overall mapping of low altitude structures to their high altitude counterparts and will provide an observation of the dynamics of the ring-current/ plasmasphere interaction.

Having demonstrated the feasibility of the ENA imaging technique to provide the capability of inferring global images of the magnetospheric energetic ion populations, we next present simulations of such images that can be expected from instrumentation designed specifically for this purpose. Figure 20 shows the result of a simulation presented by Roelof and Williams (1988) in which ENA images expected to be observed by presently designed ENA cameras (McEntire and Mitchell, 1988) are calculated from a modeled ring current injection scenario. Panels d, e, and f show the resulting ENA images as observed from a position on the dusk meridian at 8 R_E and 30° N geomagnetic latitude. These three panels represent a time sequence spanning the main phase asymmetric injection of the ring current (panel d) to the nearly symmetric recovery phase (panel f). Panels a, b, and c show the corresponding ring current ion distributions in terms of the column-integrated flux of ions whose instantaneous velocity vector points towards the viewing position. This simulation uses ring current intensities, pitch angle distributions, and their evolution from main to recovery phase that are based on available observations, a hydrogen geocoronal distribution based on DE-1 observations (Rairden et al., 1986), and a realistic magnetic field configuration for the altitudes of interest. Thus Figure 20 is a useful guide to what can be expected from the ENA imaging technique, although actual ENA images may be quite different. Note however, that the images shown in Figure 20 are similar to and consistent with those measured by the ISEE 1 MEPI (Roelof, 1987).

Roelof and Williams (1988) further present a high altitude polar orbit scenario in which they provide simulated views of the ring-current image shown in Figure 19a as it would appear as viewed from different positions in the chosen polar orbit. These results, along with those shown in Figure 20, demonstrate the potential power of the ENA technique to obtain global images of the magnetospheric energetic ion populations from many different spatial perspectives. New, important quantitative analyses will be possible using such images. For example, global images, such as those shown in Figure 20, provide a measure of the pressure tensor for the ion distribution generating the ENA image. Knowing the pressure tensor allows the currents generated by this distribution, both perpendicular and parallel to the magnetic field, to be determined (Vasyliunas, 1970). Roelof (1989) has used ISEE ENA images to obtain an estimate of the overall ring current system that includes the familiar westward current plus a radial current that agrees with statistical results based on local magnetic measurements (Iijima et al., 1988) and a parallel current that appears to connect with the auroral region II current system. These preliminary results indicate that future ENA images will have a strong quantitative impact in mapping magnetospheric current systems.

5. SUMMARY

We have discussed the need for global observations and have presented examples of present and projected capabilities of obtaining global/macroscale observations in the upper atmosphere, the ionosphere, and the magnetosphere. The addition of the previously "invisible" magnetosphere to our global observational capabilities places us in a unique position in the study of cosmological plasma environments and we end our discussion of global observations by describing this unique research position.

In the study of astrophysical systems, researchers must rely exclusively on remote sensing of the objects of interest. Obviously, there always has been a complete absence of *in-situ* measurements. As a result, astrophysicists often have good indications of the global workings of the objects of interest, but the detailed physical mechanisms that moderate the global workings must be inferred or described in theory.

In the study of magnetospheres, the situation is reversed. Detailed *in-situ* measurements have been made of the earth's magnetosphere and are proceeding within other planetary magnetospheres. As a result, we are developing a good understanding of the local physical mechanisms operating within these systems. However, we are at the very beginning of applying global imaging techniques to the magnetospheric system as a whole. As a result, we have a poor understanding of how the local physical mechanisms work together to define the overall behavior of the system. The capability of imaging magnetospheric plasmas marks the first opportunity where an astrophysical plasma system that has been well characterized by *in-situ* measurements (i.e. the Earth's magnetosphere) also can be characterized globally by remote sensing. When implemented, such a program certainly will provide a quantum jump in our understanding of the behavior of complex astrophysical plasma systems.

Figure 21 illustrates this unique opportunity of having both localized *in-situ* and global observations of an astrophysical plasma, the magnetosphere. The ENA images are taken from the ring current image inferred for the 29 September, 1978 magnetic storm (Figure 19a). Both a global and an ionospheric close-up view are shown. The ENA images are simulated for the DE-1 position (3.66 R_E altitude, 57.9° geographic latitude, and 3.7 hours geographic local time) from which the auroral VUV image was taken at 0245 UT on 13 December 1981. The ENA close-up view presents the same aspect as the auroral image. An altitude profile of ring current energy densities for various ion species (Gloeckler et al., 1985) is shown as representative of local *in-situ* observations.

The combination of both *in-situ* observations and the global observations discussed in this paper will provide an entirely new perspective of our view of the Earth's magnetosphere and its dynamics. It may, in fact, have the same revolutionary effect on our concept of such magnetized plasma systems as Prof. Van Allen's discovery of the radiation belts had on our then existing views of the Earth's space environment.

ACKNOWLEDGEMENTS

I wish to acknowledge and thank S.I. Akasofu, A.L. Broadfoot, J.R. Dudney, L.A. Frank, R.A. Greenwald, D.G. Mitchell, E.C. Roelof, and T. Rosenberg for discussions, figures, and the use of data and information prior to publication. This effort has been sponsored in part by a NASA contract to The Johns Hopkins University Applied Physics Laboratory and the Department of the Navy under Task IAGX7XXX of Navy Contract N00039-87-5301.

REFERENCES

Akasofu, S.-I., 1965, Dynamic morphology of Auroras, *Space Sci. Rev.*, IV-4:498.

Akasofu, S.-I., and S. Chapman, 1972, Solar-Terrestrial Physics, Oxford University Press, Oxford.

Baker, K.B., R.A. Greenwald, J.M. Rushoniemi, J.R. Dudeney, M. Pinnock, N. Mattin, and J.M. Leonard, 1989, PACE; Polar Anglo-American Experiment, EOS, 70:785.

Broadfoot, A.L., 1986, Images of the magnetosphere and atmosphere: global effects (IMAGE), proposal for NASA Explorer Mission Concept Studies, by D.J. Williams, L.A. Frank, A.L. Broadfoot, W.L. Imhoff, S.B. Mende, D.M. Hunten, R.G. Roble, and G.S. Siscoe.

Frank, L.A., J.D. Craven, K.L. Ackerson, M.R. English, R.H. Eather, and R.L. Carovillano, 1981, Global auroral imaging instrumentation for the Dynamics Explorer Mission, *Space Sci. Instr.*, 5:369.

Gloeckler, G., B. Wilken, W. Studemann, F.M. Ipavich, D. Hovestadt, D.C. Hamilton, and G. Kremser, 1985, First composition measurements of the bulk of the storm time ring current (1 to 300 keV/e) with AMPTE-CCE, Geophys. Res. Lett., 12:325.

Gurnett, D.A., and U.S. Inan, 1988, Plasma wave observations with the Dynamics Explorer 1 spacecraft, *Rev. of Geophys.*, 88:285.

Horwitz, J.L., R.H. Comfort, and C.R. Chappell, 1984, Thermal ion composition measurements of the formation of the new outer plasmasphere and double plasmapause during storm recovery phase, *Geophys. Res. Lett.*, 11:701.

Huff, R.L., W. Calvert, J.D. Craven, L.A. Frank, and D.A. Gurnett, 1988, Mapping of auroral kilometric radiation sources to the aurora, *J. Geophys. Res.*, 10:11445.

Iijima, T., T.A. Potemra, and L.J. Zanetti, 1988, Large-scale characteristics of magnetospheric equatorial currents, *JHU Applied Physics Laboratory Preprint*, 88-14.

Keath, E.P., G.B. Andrews, A.F. Cheng, S.M. Krimigis, B.H. Mauk, D.G. Mitchell, and D.J. Williams, 1989, Instrumentation for energetic neutral atom imaging of magnetospheres, in Waite et al. (Eds.), Solar System Plasma Physics, Geophysical Monograph 54, American Geophysical Union, Washington D.C., p. 165.

McEntire, R.W., and D.G. Mitchell, 1989, Instrumentation for global magnetospheric imaging via energetic neutral atoms, in Waite et al. (Eds.), Solar System Plasma Physics, Geophysical Monograph 54, American Geophysical Union, Washington D.C., p. 69.

Newell, H.E., 1959, Capabilities for space research, *J. Geophys. Res.*, 4:1695.

Rairden, R.L., L.A. Frank, and J.D. Cruven, 1986, Geocoronal imaging with Dynamics Explorer, *J. Geophys. Res.*, 91:13613.

Roelof, E.C., D.G. Mitchell, and D.J. Williams, 1985, Energetic neutral atoms (E~50 keV) from the ring current: IMP 7/8 and ISEE 1, *J. Geophys. Res.*, 90:10991.

Roelof, E.C., 1987, Energetic neutral atom image of a storm-time ring current, *Geophys. Res. Letters*, 14:652.

Roelof, E.C., and D.J. Williams, 1988, The terrestrial ring current: From *in-situ* measurements to global images using energetic neutral atoms, *Johns Hopkins APL Technical Digest*, 9:144.

Roelof, E.C., 1989, Remote sensing of the ring current using energetic neutral atoms, in press, *Advances in Space Research*, 1989.

Swift, D.W., R.W. Smith, and S.-I. Akasofu, 1989, Imaging the Earth's magnetosphere, *Planet. Space Sci.*, 37:379.

Van Allen, J.A., G.H. Ludwig, E.C. Ray, and C.E. McIlwain, 1958, Observation of high intensity radiation by satellites, *Jet Propul.* 28:588.

Van Allen, J.A., C.E. McIlwain, and G.H. Ludwig, 1959, *J. Geophys.Res.*, 64:271.

Van Allen, J.A., and L.A. Frank, 1959, Radiation around the Earth to a radial distance of 107,000 kilometers, *Nature*, 183:430.

Vasyliunas, V.M., 1970, Mathematical models of magnetospheric convection and its coupling to the ionosphere, Particles and Fields in the Magnetosphere, Ed. B.M. McCormac, D. Reidel Pub. Co., p. 60.

Waite, J.H., J.L. Horwitz, and R.H. Comfort, 1984, Diffusive equilibrium distributions of He^+ in the plasmasphere, *Planet. Space Sci.*, 32:611.

ROLE OF SMALL SATELLITE MISSIONS IN MAGNETOSPHERIC RESEARCH

Stanley D. Shawhan

Space Physics Division, Code ES
Office of Space Science and Applications
NASA Headquarters
Washington DC 20546 USA

ABSTRACT

Since the beginning of space research in 1957, more than 100 small scientific satellites have been launched to explore the region surrounding the earth and to characterize this region which is known as the "magnetosphere". A primary contributor to this research effort has been Dr. James A. Van Allen, the winner of the 1989 Crafoord Prize.

1. EARLY MAGNETOSPHERIC RESEARCH

On October 4, 1957, Sputnik 1 was successfully launched by the Soviet Union ushering in an era of space research. On February 1, 1958 the Explorer 1 satellite was launched by the United States making measurements of the earth's upper atmosphere which indicated the presence of energetic particles and magnetic fields. During the next several years series of small research satellites from the Soviet Union—Sputnik, Luna and Kosmos—and from the United States—Explorer and Pioneer—expanded our knowledge of "space" at increasing distances from the earth.

During the early 1960's, the picture quickly emerged of the "magnetosphere", the region surrounding the earth which is formed by the extension of the earth's magnetic field into space. The first real discovery of the space age was the presence of trapped energetic particles within the magnetosphere by Van Allen and his colleagues; these regions have become known as the Van Allen Radiation Belts (see artist's concept in Figure 1). This discovery was significant for two reasons. It was known that the earth was bombarded by cosmic rays, but, theorists could not agree whether cosmic ray energetic particles could be trapped in the earth's field; the first measurements showed that they could. Secondly, for those planning manned space missions, the radiation hazard to the crew members had to be taken into account.

Close to the earth, the magnetic field appeared to be a simple dipole, as expected. Far from the earth, it was discovered that the field was compressed on the dayside to about 20 earth radii (about 120,000 km) and extended on the nightside well beyond the orbit of the moon at

Magnetospheric Physics, Edited by B. Hultqvist and C.-G. Fälthammar
Plenum Press, New York, 1990

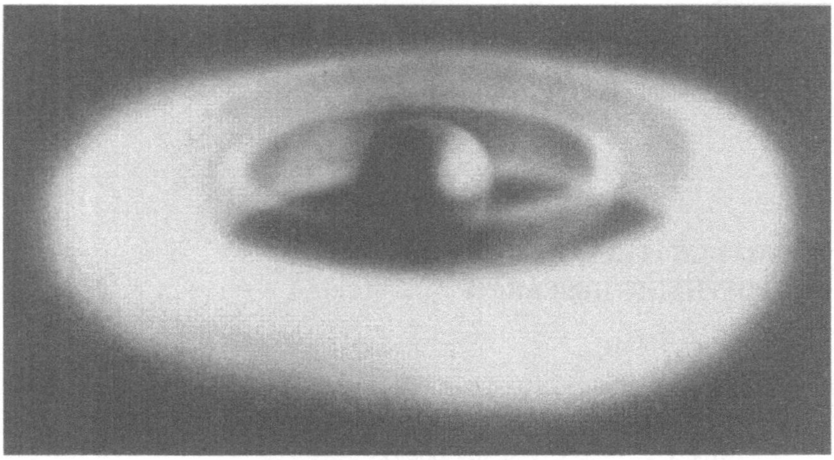

Fig. 1. Artist's concept of the Van Allen radiation belts surrounding the Earth (courtesy of the University of Iowa).

60 earth radii (380,000 km) due to the influence of a "wind" of magnetic fields and charged particles continually blowing from the sun throughout interplanetary space. It was discovered that this solar wind created a shock wave in front of the magnetosphere as well as compressing the magnetic fields. These and other discoveries about the earth's magnetosphere are covered in the paper by Hultqvist (1989), entitled "Major Achievements of Magnetospheric Research".

The early explorations of the magnetosphere have evolved into detailed, focused, studies of many different phenomena with the goal of understanding of the magnetosphere as a system. Models are now being developed that can describe cause-and-effect relationships, can serve as predictors of solar and human influence and can couple to models of the lower atmosphere for studying global change. Some of the science topics that have been included in magnetospheric research over the past 32 years include the following:

- solar radiation effects on the atmosphere
- solar wind - bow shock interactions
- magnetopause structure and dynamics
- active plasma injection experiments
- ionospheric instabilities and turbulence
- lower thermospheric dynamics and coupling
- chemistry and dynamics of the upper atmosphere
- stimulation of space plasma processes through wave and particle injection experiments

2. SMALL MAGNETOSPHERIC SATELLITE PROGRAMS

Many nations have developed satellite programs for magnetospheric research. Compared to manned programs and space applications such as weather and communications satellites, these research satellites are small. Masses range from 5 kg for Explorer 1 up to 600 kg for Interkosmos satellites. The characteristic size of the spacecraft is 1 meter, which houses most of the particle sensors, with appropriate appendages such as 5 meter booms for sensors, like magnetometers, and 100 meter plasma wave antennas. Program costs have increased to more than $100 million per year for the U.S. National Aeronautics and Space Administration (NASA) Explorer Program, the European Space Agency (ESA) "Blue" or "Yellow" missions,

Table 1

• Successful Space Launches 1957 - 1987 Worldwide	2979	
• Scientific Satellites Terrestrial, Planetary, Solar Wind, Astrophysics Not Operational--NOAA, GOES, DMSP, METEOR	267	9%
• Magnetospheric Satellites Not Observatories--OGO, Some Kosmos, Prognoz, ATS or OV	178	67%
• Small Magnetospheric Satellites Sputniks, Explorers, Elektrons, IMPs, Alouette/ISIS, ISAS-Series, Ariels, Interkosmos, San Marco, Viking	137	77%

*TRW Space Log 1957-1987, 1988, Vol 23

and the on-going Japanese Institute of Space and Astronautics Science (ISAS) program with one launch per year; the Soviet Space Research Institute (IKI) is just beginning their own "Regatta" small magnetospheric satellite program.

Within the major space agencies—NASA, ESA, ISAS and IKI—as well in other national programs, the role and goals of small magnetospheric research programs are very similar (Committee on Solar and Space Physics, 1984):

- primary tool for detailed study of particular regions and phenomena
- cost-effective method of space exploration
- flight demonstration of new spacecraft and instrument designs
- bridge gap between sounding rockets and major space observatories
- quick reaction for space science targets of opportunity
- training opportunities for students and young scientists
- basis for international cooperative missions
- frequent flight opportunities

Table 1 "Space Mission Statistics" gives the number and percentage of launches in different categories. Using the TRW Space Log 1957-1987 (1988) as the reference but adding the scientific satellite launches since 1987 by ISAS (EXOS-D, also known as Akebono), ESA (Hipparcos) and IKI (Activnyi), there have been approximately 267 scientific satellites launched or 9% of all launches (approximately 3000). Of these scientific satellites, 178 or 67% have investigated the earth's magnetosphere including large observatories. Within the small magnetospheric class are 137 satellites or 77% of all magnetospheric missions and 51% of all scientific satellites. As tabulated in Table 2 "Small Magnetospheric Satellites by Nation", 95% of the small magnetospheric missions have been under the direction of the U.S., the USSR, Europe and Japan with 5% by Canada, India, the Peoples Republic of China and Czechoslovakia.

Table 2

• United States NASA (48) and Other US Government (20)	68	50%
• Soviet Union All Agencies	28	20%
• Europe UK (7), ESRO/ESA (6), Italy (5), France (4), Germany (3), Sweden (1)	26	19%
• Japan ISAS & NASDA	8	6%
• Canada	4	3%
• Other India, PRC & Czechoslovakia	3	2%

*TRW Space Log 1957-1987, 1988, Vol 23

At the start of the space era there was a steady increase in the number of small magnetospheric satellites launched and in the mass of each satellite as shown in Figure 2 "Time History of Small Satellites". These increases in numbers and mass were probably keyed to the increases in launch capability of the Soviet Union and the United States. In the mid to late 1960's the earth's vicinity was well explored. By that time, larger spacecraft were needed to make many measurements simultaneously to obtain cause-and-effect relationships; also, the launch capability became such that large, longer-lived, observatories could be launched and operated. Therefore, the launch rate of small satellites decreased. Also in the 1970's, with limited space budgets, more scientific disciplines began to make use of space for probes to other planets, for astrophysics and solar telescopes and for earth observations. Therefore during the 1970's and 1980's there have been fewer space missions dedicated to the original space science discipline—magnetospheric physics. In the period of 1985 to the present, there is only about one launch per year for magnetospheric research; the magnetospheric research community feels that this rate is not sufficient to provide the community with new data or to progress at a reasonable pace to the stage of understanding the magnetospheric system.

One of the goals identified for the small magnetospheric satellite programs is to provide a basis for international cooperative programs. The international nature of magnetospheric research is reflected in Table 2; missions have been deployed by more that 12 nations plus the ESA/ESRO consortium.

Perhaps the best example of an international mission is the Active Magnetospheric Particle Tracer Explorers (AMPTE) mission. Three satellites were launched together on August 16, 1984 by a U.S. Delta rocket. The Ion Release Module (IRM) was provided by the Federal Republic of Germany, the United Kingdom Satellite (UKS) by the U.K. and the Charge Composition Explorer (CCE) by the U.S. The three satellites are shown in Figure 3 (Acuna, et al.,1985). In the case of AMPTE, multipoint measurements were required to support chemical release experiments in the solar wind and in the tail of the magnetosphere. The required instrument sets and the satellites were possible only through the contributions of the three

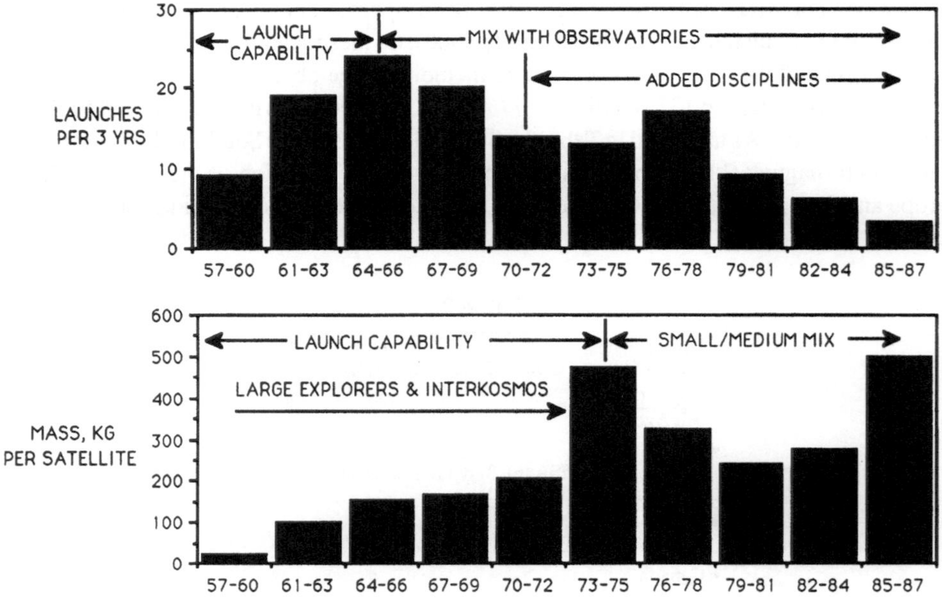

Fig. 2. Time history of small satellites for magnetospheric research (based on data from the TRW Space Log 1957-1987, 1988).

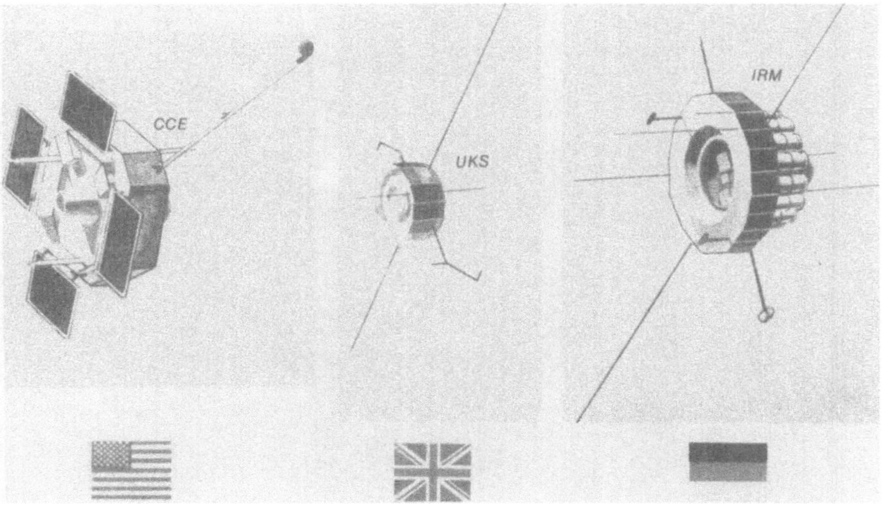

Fig. 3. AMPTE satellite set from the U.S., the U.K. and the FRG (NASA, 1984).

nations. Likewise the scope of the science problems being addressed required cooperation between the science communities.

Roederer (1989) describes past, present and future programs where international cooperation has been established not only for spaceflight missions, but, also between the groundbased, rocket, balloon and spaceflight investigators with the goal of obtaining a larger science community and a more comprehensive level of understanding.

3. INNOVATION

As the small magnetospheric satellite programs have evolved, they have been characterized by innovative developments in instrumentation, orbits, science and satellite engineering.

Besides the new development of the orbiting satellite itself in 1957, one instrument system was essential to the discovery of the radiation belts—the tape recorder. In Figure 4 is shown the tape recorder planned for Explorer 1, but, dropped due to weight limitations; it was finally flown successfully on Explorer 3. This 5 cm (2 inch) tape recorder was designed and built by Dr. Van Allen and his colleagues at the University of Iowa. The tape recorder had the capability to record data from the Geiger tube radiation detectors for a full orbit. On playing back the data, orbit after orbit, it was possible to see the pattern that led to the conclusion that energetic particles were trapped in the earth's magnetic fields in several different zones.

Orbits of the first satellites were nearly circular, just above the atmosphere. As the launch capability increased, satellites were sent far from the earth in highly elliptical orbits. In October 1958 Pioneer 1 reached 70,000 km apogee, Luna 1 escaped earth orbit after passing close to the moon in January 1960 and the planets Venus and Mars were reached in the mid 1960's by U.S. Pioneers and the Soviet Venera and Mars series.

However, the most impressive display of orbital mechanics was accomplished by the International Sun-Earth Explorer 3 (ISEE 3). In August 1978, ISEE 3 was the first satellite launched and inserted into an orbit about the L-1 sun-earth libration point. The L-1 point is located 1.5 million km on the dayside of the earth; ISEE 3 was not orbiting about the earth but rather at the L-1 point. At this location it monitored changes in the solar wind in support of its

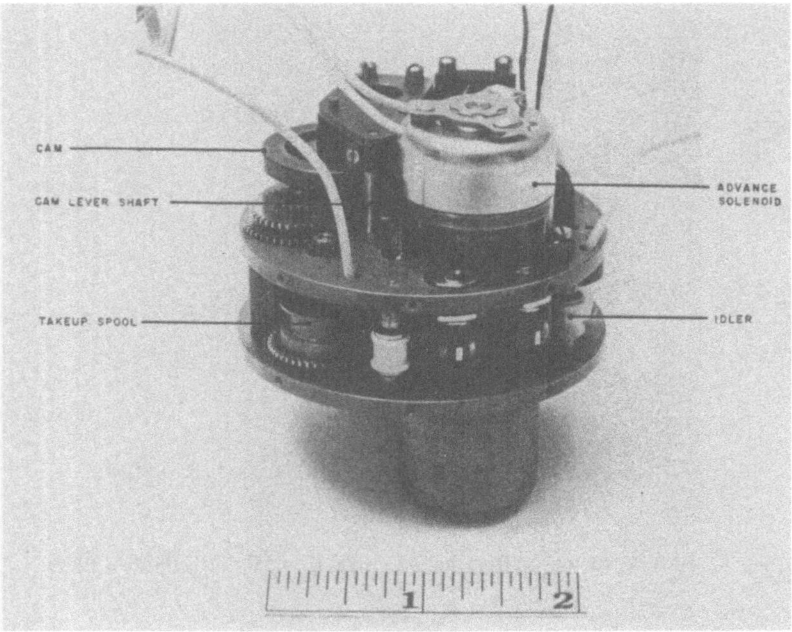

Fig. 4. Photograph of the Explorer 3 tape recorder which enabled the discovery of the radiation belts (courtesy of the University of Iowa).

sisters, the earth co-orbiting pair of spacecraft NASA's ISEE 1 and ESA's ISEE 2. In June 1982, the spacecraft was diverted to a second mission by a close encounter with the moon to orbits through the geomagnetic tail. These two deep tail orbits were the first systematic explorations of the nightside magnetosphere to 1.5 million km. Finally in December 1983, ISEE 3 was diverted to a third mission again by a series of lunar swingby maneuvers; the last maneuver skimmed past the moon at a scant 120 km altitude. For this third mission, the satellite was renamed the International Cometary Explorer (ICE); it became the first satellite to fly close to a comet and in fact penetrated the tail region of comet Giacobini-Zinner on September 11, 1985. Currently, ICE continues to provide solar wind data at 1 AU from the sun but at increasing earth-sun-ICE angles. It is expected to return to the vicinity of the earth in 2012. The convoluted trajectory of ISEE 3/ICE is shown in Figure 5.

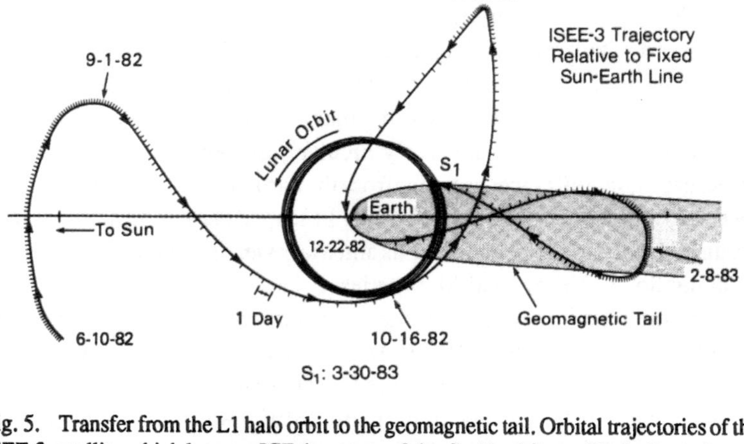

Fig. 5. Transfer from the L1 halo orbit to the geomagnetic tail. Orbital trajectories of the ISEE 3 satellite which became ICE (courtesy of the Goddard Space Flight Center).

The AMPTE mission has already been cited above for its significant international involvement. AMPTE is also representative of innovative science investigations within small magnetospheric missions. The primary goal of the AMPTE mission was to investigate plasma expansion, entry into the magnetosphere and energization through diffusion processes through release and tracing of lithium and barium ions. Figure 6 depicts the orbits of the IRM/UKS co-orbiting pair covering both the solar wind and the magnetospheric tail and the CCE within the equatorial ring current region. Ion tracer releases are indicated in the solar wind just in front of the bow shock to study plasma entry into the magnetosphere, in the magnetosheath region to produce an artificial comet and in the magnetospheric tail to investigate plasma diffusion and energization processes (Acuna, et al.,1985). This 6 month period of active chemical release investigations was unique in that theories of plasma transport which were developed from 25 years of exploration and discovery were tested using new innovative techniques. Besides these active experiments, AMPTE also returned the first measurements of the elemental and charge composition and dynamics of the charged particle population in the magnetosphere over a wide energy range. These measurements by the CCE confirmed the hypothesis that charged oxygen must be a significant component of the ring current region, for example.

Other examples of active experiments that have yielded significant results for the understanding of the magnetosphere are reported by Van Allen (1989).

After more than 100 small magnetospheric research satellites one would think that new missions would just utilize the "tried and true" designs. Now that the Soviet Space Research Institute in Moscow has received approval to develop a new small satellite series it is taking a new approach in a number of areas. A sketch of the "Regatta" small scientific laboratory is shown in Figure 7. The obvious innovation in this design is the inclusion of a "solar sail" to provide passive pointing of the spacecraft to the sun and the possibility of some orbit change.

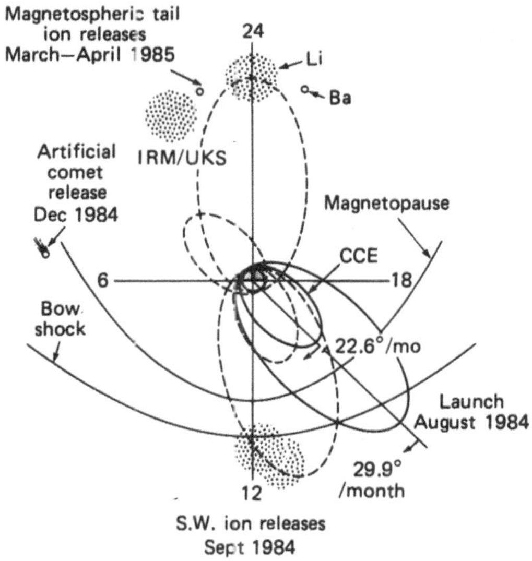

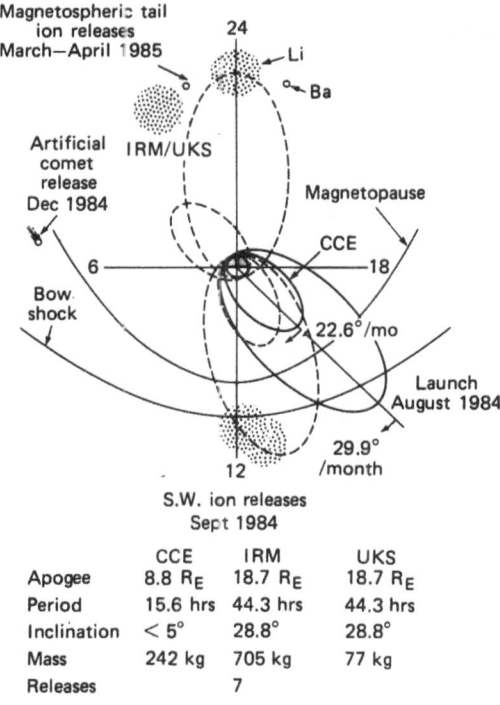

	CCE	IRM	UKS
Apogee	8.8 R_E	18.7 R_E	18.7 R_E
Period	15.6 hrs	44.3 hrs	44.3 hrs
Inclination	< 5°	28.8°	28.8°
Mass	242 kg	705 kg	77 kg
Releases		7	

Fig. 6. AMPTE satellite orbits and chemical release locations (Acuna, 1985).

With this scheme, it may be possible to avoid use of an attitude control system and all the attendant mass and complexity. Another innovation is to make use of carbon composite structure elements. The sail and structure together could allow accommodation of a payload of 250 kg at 50% of the total mass (compared to conventional payloads of 20% mass). Because the spacecraft does not spin, the design includes an angular momentum compensated spin platform for instruments that must scan for 3-D coverage. Also, GaAs solar concentrator arrays are to be used for power where each instrument is provided a dedicated string of cells in hopes of minimizing electromagnetic interference between instruments and subsystems. Two magnetospheric Regatta missions are being developed: one for launch in late 1993 into an equatorial orbit at 2 by 12 earth radii to complement the NASA Global Geospace Science missions and one in late 1995 into polar orbit at 2 by 20 earth radii to complement the ESA Cluster mission (Reinhard, 1987). Two others are in planning for 1996-1998 at the sun-earth libration point and/or the trans-lunar region.

Other new mission concepts which make use of multiple satellites have been put forward by Haerendel (1989) in these proceedings.

4. RENEWAL OF THE SMALL MISSION CONCEPT

According to Figure 2, the number of small magnetospheric satellite missions reached a peak of 6 per year in the mid-1960's but has decreased to the current level of about 1 per year worldwide. There are several factors which portend a reversal of this downward trend: the launch capability for small satellites is multiplying across several nations; designs for small satellites are relatively simple and have been somewhat standardized; and the science community has identified many research questions that could be answered with several smaller, focused, missions.

Creation of the Soviet Regatta program by IKI was mentioned above. This program gives IKI the flexibility to develop and control their own space missions at a modest cost but in a way

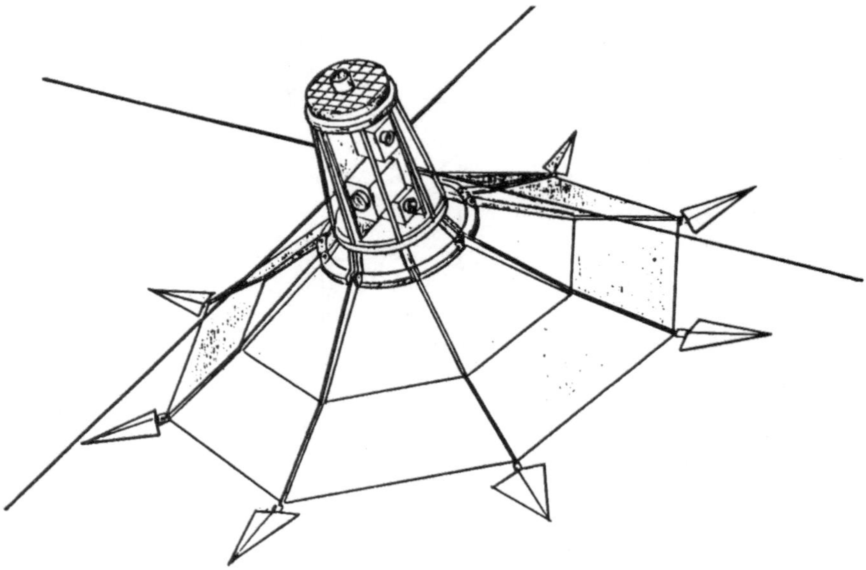

Fig. 7. Soviet "Regatta" spacecraft concept (courtesy of the Space Research Institute, Moscow).

Table 3

• Solar, Anomalous & Magnetospheric Particle Explorer (SAMPEX)

= Cosmic & Heliospheric Physics	- solar energetic particles
(uses geomagnetic cutoff in LEO)	- interplanetary particles
	- galactic cosmic rays
	- precipitating electrons

= 580 km, 82 deg inclination (WTR); 3 year lifetime
= June 1992 (target); Scout launch vehicle
= U of MD, MPI, Cal Tech, GSFC, Aerospace

• Fast Auroral Snapshot Explorer (FAST)

= Space Plasma Physics	- auroral processes
(high time resolution)	- plasma flows
	- magnetic & electric fields

= 300 x 3500 km, 76 deg inclination (WTR); 2 year lifetime
= December 1993 (target); Scout launch vehicle
= UC Berkeley, LPARL, UC Los Angeles

that takes advantage of the significant Soviet launch capability. According to their plans, they should be able to launch a satellite every other year. Even smaller nations can deploy impressive missions; for example, the Swedish "Viking" mission which was launched in February 1986 to make the first really high time resolution measurements of fields and particles in auroral regions. Now Sweden and the Federal Republic of Germany are teaming to deploy a follow-on auroral mission called "Freja". A third mission named IMPACT is also discussed. It may include two satellites making use of an electron beam to map the magnetic and electric fields between the satellites.

In Table 2, the U.S. is listed as having launched about half of the small magnetospheric missions; at the present time, however, the launch rate has fallen to one every several years. In fact, there has not been a launch of this type since the 1984 AMPTE mission (although the Italian/U.S. San Marco mission was launched in March 1988). In 1988, however, NASA re-instituted the "Small Explorer Program" (SMEX Program) with an emphasis on lower cost, more frequent missions for focused science objectives in the Space Physics, Astrophysics and Upper Atmospheric disciplines. More than 50 proposals were received from which 4 were selected for flight. Data on the two missions which fall into the magnetospheric category are given in Table 3; the "Solar, Anomalous and Magnetospheric Particle Explorer" (SAMPEX) is scheduled for June 1992 launch and the "Fast Auroral Snapshot Explorer" (FAST) for December 1993. The goal of the SMEX Program is for a launch at least every other year. Mozer (1989) describes the virtues of several missions including Viking and FAST in these proceedings .

Although there have been more than 100 missions to study the magnetosphere, the research has not reached the level of sufficient understanding where models can quantitatively predict effects; many processes and large scale features require more detailed measurements, theory (see Kennel, 1989) and simulations (see Ashour-Abdalla, 1989). Many topics for investigations have been identified:

• auroral and magnetospheric imaging
• magnetic fieldline tracing via chemical releases
• magnetosphere-ionosphere-atmosphere coupling
• ionospheric turbulence and electromagnetic coupling
• auroral electrodynamics
• non-linear wave - particle interactions and acceleration
• thermospheric response to solar irradiance
• imaging of Io plasma torus
• dust - plasma interactions

Some of these topics are addressed in detail in the paper of Williams (1989) in this volume. With current technology, small magnetospheric missions may be extended to other planets; goals for research at other planets have been identified and discussed during this symposium by Lanzerotti (1989).

5. CONTRIBUTIONS OF J. A. VAN ALLEN

The 1989 Crafoord Prize winner, Prof. James A. Van Allen is best known for the discovery of the radiation belts surrounding the earth. However, to this day he continues to contribute to many other discoveries such as the extent of the magnetosphere, including the radiation belts, at the earth and at other planets—Jupiter and Saturn—and the lack of magnetospheres at Venus and Mars. He helped to characterize the solar wind, solar energetic particles, and solar x-rays, for example. Now he continues to estimate the extent of the "heliosphere" from solar and galactic cosmic ray measurements on Pioneers 10 and 11 as they progress toward the boundary between the extent of the solar magnetic field and interstellar space.

Prof. Van Allen has participated in the following 29 U.S. science missions as a Principal Investigator which have led to the many significant scientific contributions, the education of research students and the development of scientific colleagues:

- Explorer 1, 3, 4, 7, 12, 14 (built at Iowa)
- Injun 1, 3, 4, 5 (built at Iowa)
- Hawkeye (built at Iowa)
- Ranger 1, 2
- Interplanetary Monitoring Platform 4, 5 ,6
- Orbiting Geophysical Observatory 1, 2, 4
- Mariner 2, 3, 4, 5, 9
- Pioneer 3, 4, 10, 11
- Galileo

The Van Allen organization at the University of Iowa has been unique in that, especially in the early years, they not only built the scientific instruments, but, also built the spacecraft ; at least a dozen spacecraft were designed and built at Iowa starting with Explorer 1 in 1958 and ending, so far, with the Plasma Diagnostics Package that was deployed and recovered by the Spacelab 2 Shuttle mission in 1985. In addition, the Iowa team commanded many of the spacecraft and acquired the data themselves, first from the antennas on the roof of the Physics Building, then later through the North Liberty Radio Observatory (Wells, 1980). The largest of the Iowa fabricated spacecraft was Injun 5, weighing 71 kg, pictured in Figure 8; it made comprehensive measurements of fields, plasma waves, thermal plasma, suprathermal plasma, and energetic particles in the auroral regions starting in August 1968. Injun 5 is characteristic of the small magnetospheric satellite class. The Hawkeye spacecraft was the smallest of the modern era with a mass of 23 kg; it attained a polar elliptical orbit in June 1974 with apogee of 125,000 km above the north pole. It made some of the first measurements of polar magnetic fields and plasma entry through the dayside cusp. Because Hawkeye could be tracked from North Liberty for 90% of its period, realtime, processed, data were available for the first time from an Explorer satellite.

Just for fun, the author imagined the NASA "peer review" response to Prof. Van Allen's original proposal for Explorer 1. Here is how the proposal might be evaluated today in terms of scientific, technical and programmatic criteria:

Fig. 8. Photograph of the Injun 5 satellite undergoing testing (courtesy of the University of Iowa).

- Scientific
  cosmic rays have already been measured
  trapped radiation is theoretically impossible

- Technical
  you do not need a satellite when you have rockoons
  those Geiger tubes will never survive launch
  no thermal model or NASTRAN analysis is proposed

- Programmatic
  cost is under-estimated: $66,000 proposed whereas
  the model predicts $55,000 per kg x 5 kg = $275,000!
  R & QA plan is inadequate
  J. A. Van Allen is <u>not</u> a qualified NASA solderer

Based on this proposal evaluation and the projected cost to NASA, this proposal would probably be rejected!

REFERENCES

Acuna, M. H., Ousley, G. W., McEntire, R. W., Bryant, D., Paschmann, G., 1985, Editorial: AMPTE-mission overview, *Geoscience and Remote Sensing,* GE-23:175.

Ashour-Abdalla, M., 1989, Roles of simulations in future magnetospheric programs, *in:* "Magnetospheric Physics: Achievements and Prospects", B. Hultqvist and C.-G. Fält-hammar, eds., Plenum Press, New York.

Committee on Solar and Space Physics, 1984, "A Strategy for the Explorer Program for Solar and Space Physics", National Academy Press, Washington.

Haerendel, G., 1989, Why do we need multipoint measurements?, *in:* "Magnetospheric Physics: Achievements and Prospects", B. Hultqvist and C.-G. Fälthammar, eds., Plenum Press, New York.

Hultqvist, B.,1989, Major achievements of magnetospheric research, *in:* "Magnetospheric Physics: Achievements and Prospects", B. Hultqvist and C.-G. Fälthammar, eds., Plenum Press, New York.

Kennel, C. F., 1989, The role of theory programs in magnetospheric research, *in:* "Magnetospheric Physics: Achievements and Prospects", B. Hultqvist and C.-G. Fälthammar, eds., Plenum Press, New York.

Lanzerotti, L. J., 1989, Why do we need missions to the magnetospheres/ionospheres of other planets?, *in:* "Magnetospheric Physics: Achievements and Prospects", B. Hultqvist and C.-G. Fälthammar, eds., Plenum Press, New York.

Mozer, F. S., 1989, Why do we need higher resolution measurements?, *in:* "Magnetospheric Physics: Achievements and Prospects", B. Hultqvist and C.-G. Fälthammar, eds., Plenum Press, New York.

NASA, (1984), "AMPTE", NASA Brochure EP-215. Washington.

Reinhard, R., 1987?, "Inter-Agency Consultative Group Missions", ESA Publications Division, Noordwijk, The Netherlands.

Roederer, J. G., 1989, International coordination of magnetospheric research, *in:* "Magnetospheric Physics: Achievements and Prospects", B. Hultqvist and C.-G. Fälthammar, eds., Plenum Press, New York.

"TRW Space Log 1957-1987", 1988, Thompson, T. D., ed., TRW Space and Technology Group, Redondo Beach CA.

Van Allen, J. A., 1989, Active experiments in magnetospheric physics, *in:* "Magnetospheric Physics: Achievements and Prospects", B. Hultqvist and C.-G. Fälthammar, eds., Plenum Press, New York.

Wells, J. P., 1980, "Annals of a University of Iowa Department: From Natural Philosophy to Physics and Astronomy", Department of Physics and Astronomy, University of Iowa, Iowa City.

Williams, D. J., 1989, Why do we need large scale imaging?, *in:* "Magnetospheric Physics: Achievements and Prospects", B. Hultqvist and C.-G. Fälthammar, eds., Plenum Press, New York.

WHY DO WE NEED GROUND BASED, BALLOON AND SOUNDING ROCKET MEASUREMENTS IN THE FUTURE?

Gordon Rostoker

Institute of Earth & Planetary Physics
and Department of Physics
University of Alberta
Edmonton, Alberta, Canada, T6G 2J1

ABSTRACT

Now that vast amounts of *in situ* data have been and continue to be gathered by satellites in the Earth's magnetosphere, it is sometimes felt that there is no longer a use for data obtained through traditional remote sensing techniques from the ground or from balloon borne probes. In this paper it will be shown that data from arrays of ground based remote sensors play an instrumental role in defining the global characteristics of magnetospheric and ionospheric activity, without which interpretation of the *in situ* data acquired by satellites is severely limited. It will also be shown that there still is a need for experiments launched aboard sounding rockets in terms of the study of ionospheric plasma and wave properties and also for the purpose of carrying payloads needed to initiate active experiments in geospace.

1. INTRODUCTION

Research in the field of magnetospheric physics began with ground observations of certain manifestations of the solar terrestrial interaction which were detectable by eye or through the use of simple instrumentation. The earliest observations were those of the aurora borealis which have been documented since the times of the Greek philosophers such as Aristotle and Anaxagoras. During the early part of the 19th century, magnetic variations were being used to infer the level of the solar terrestrial interaction, and by the beginning of the 20th century the relationship between auroras and magnetic perturbations detected at the Earth's surface was well recognized (cf. Birkeland, 1908; Chree, 1909). As tools to measure the high frequency portion of the electromagnetic spectrum were developed, the rich complexity of the signatures of magnetospheric activity became evident. Ground observations were used effectively to infer certain characteristics of the sun-Earth system and the properties of geospace. Obvious examples of major achievements in this respect were the discovery of the continuous nature of the solar wind using observations of comet tails (Biermann, 1951), the existence and structure of the ionosphere using radio waves (Appleton and Barnett, 1925) and the number density of magnetospheric plasma in the equatorial plane together with the existence of the plasmapause using whistler data (Storey, 1953; Carpenter, 1963).

By the 1950's, rockets were being used to carry payloads to altitudes where radiation detectors could monitor the bremsstrahlung X-rays caused by the encounters of primary energetic electron precipitation with atoms and molecules of the upper atmosphere (Meredith et al., 1955). In addition, balloon borne detectors were also able to monitor the X-ray fluxes and by the 1960's researchers such as Barcus and Brown (1966) had been able to relate the observed X-rays to auroral luminosity and to ionospheric absorption as measured by riometers.

Since the advent of *in situ* measurements of particles and fields in the ionosphere and magnetosphere, it has been possible to measure many of the key physical parameters needed to define the physics of the solar terrestrial interaction. Since the measurements can be made directly using instrumentation carried aboard orbiting satellites, it seems reasonable to consider the possibility that it is no longer necessary to infer the plasma and field properties of the magnetosphere-ionosphere system from remote sensing measurements made using instrumentation located below the ionosphere. The purpose of this paper is to show that now, more than ever, the remote sensing techniques are required to establish the large scale properties of the global plasma and field configuration that are not realistically open to *in situ* sampling with an adequate density of observation points.

Perhaps the most effective way of highlighting the importance of remote sensing measurements using detectors sited below the ionosphere is to point out that there is a general paucity of simultaneous data acquired at different locations in the magnetosphere. Most studies in the past have involved data from a single satellite and have featured statistical properties of the data sets. It should be clear that, having data from a few arbitrarily placed satellites in a volume of $\sim 10^{12}$ km^3 involving a complex configuration of plasmas and fields is totally inadequate if one is to acquire a reasonably unique picture of the solar terrestrial interaction. Fortunately, one end of the volume of plasma represented by the magnetosphere-ionosphere system is the ionosphere itself. Remote sensing techniques permit the electrodynamic and plasma properties of the ionosphere to be defined in global terms. It should therefore be clear that information relevant to the surface separating the magnetosphere from the neutral atmosphere can provide important constraints on any physical model proposed to quantitatively describe magnetosphere-ionosphere coupling.

One can also appreciate the value of estimates of ionospheric properties based on data from sensors located on the ground or in the neutral atmosphere by realizing that detectors aboard a single satellite can only yield information at one point along any field line. If one can describe the properties of the ionosphere at all points on the interface region between the neutral atmosphere and the magnetosphere, one is guaranteed a second point of observation on the field line passing through the satellite. Evidently, having data from two observation points on a field line provides the researcher with considerably more information on which to base a description of the physical processes taking place along that field line. There is, however, a problem in taking this approach. Despite considerable efforts in building models of the geomagnetic field in the past (cf. Stern, 1987), we still do not have a good model for mapping the Earth's magnetic field from the ionosphere into the distant magnetosphere. The modern models constructed by Tsyganenko (1987) and Hau et al. (1989) have provided researchers with some insight into the modeling problem together with rudimentary user-friendly algorithms for computing the magnetic field of the Earth. However, these models still lack important components (such as field-aligned currents) making mappings developed using the algorithms somewhat suspect. Developing a fully acceptable magnetic field model of the Earth's magnetosphere remains one of the most important problems in magnetospheric physics and one which must be successfully dealt with before one can effectively use remote sensing information about ionosphere properties to complement data obtained using orbiting satellites.

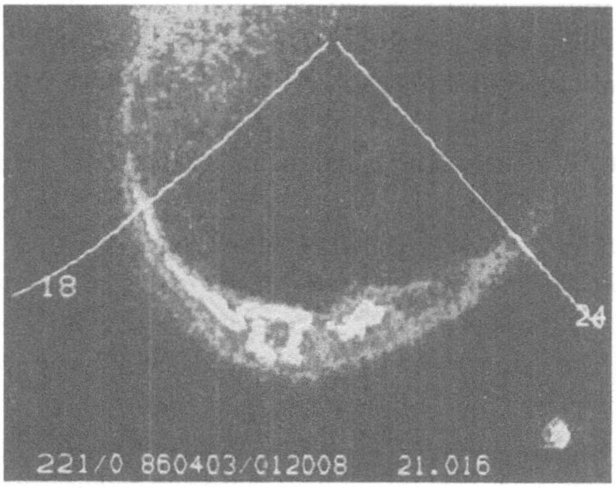

Fig. 1. Snapshot of the late afternoon and evening sector of the auroral oval taken by the Viking imager at 0120:08 UT on April 3, 1986. It is clear that regions of high luminosity show strong spatial localization even in the central plasma sheet well equatorward of the poleward edge of the oval. Simple averaged models of auroral oval conductivity are evidently not easily applicable to the study of individual events when it comes to quantitative modeling.

Finally, it is important to note that trying to understand the nature of the solar terrestrial interaction represents a global problem which requires simultaneous knowledge of the plasma and field parameters throughout the magnetosphere-ionosphere system. Such information is not going to be available to space researchers in the foreseeable future using only data from orbiting satellites. It is absolutely imperative to supplement the available satellite data with information about the plasma and field properties at the feet of field lines threading the magnetosphere. The balance of this paper will be devoted to outlining the type of information which can be obtained from remote sensing techniques using ground based instrumentation along with data acquired using balloon borne detectors. In addition, the value of data acquired using rocket borne detectors in the context of modern space physics will also be discussed.

2. GROUND BASED REMOTE SENSING TOOLS

If one inspects the past literature in which satellite data are correlated with auroral oval activity levels, one finds that in many cases little attention is paid to where the satellite is located in local time. Failing to incorporate this information into a correlative study is tantamount to assuming that the auroral oval is structureless (i.e. no matter where in the auroral oval the foot of the field line threading the satellite is found, the ionospheric plasma and field parameters will be the same). Figure 1 shows a picture of the evening sector auroral oval taken by the Viking satellite's UV imager during the course of a large magnetospheric substorm. The spatially localized auroral structures along the poleward edge of the oval together with the dramatic north-south aligned structures in the equatorward portion of the oval clearly demonstrate that the plasma and field characteristics of the nightside auroral ionosphere are far from uniform. Researchers are therefore interested in obtaining some indication of the position of localized structures in the oval so that they can better interpret data acquired by their satellite borne detectors. The primary tools employed by the ground based research community are:

(1) Magnetometers used to infer information about the electric current systems which couple the magnetosphere and ionosphere.

(2) Riometers used to infer the pattern of energetic electron precipitation in the high latitude ionosphere.

(3) All-sky cameras and TV imagers used to follow rapid variations in the fine auroral structure of auroras and meridian scanning photometers used to define the gross auroral structure in the local field of view.

(4) Coherent scatter and incoherent scatter radars used to infer information about auroral electric fields and plasma number densities and temperatures.

(5) Interferometers used to define the neutral wind velocity field.

Over the past two decades these techniques have been developed to the point where they are powerful tools for inferring a wide range of ionospheric properties, as will be shown in the following sections.

2.1. *Magnetometers*

Figure 2 shows a typical magnetogram from an auroral oval station (covering a period of substorm activity during which time the station sampled the hours around magnetic midnight). Prior to the development of coordinated arrays of magnetometers in the late 1960's, it was customary to try to establish the level of magnetospheric activity using single magneto-

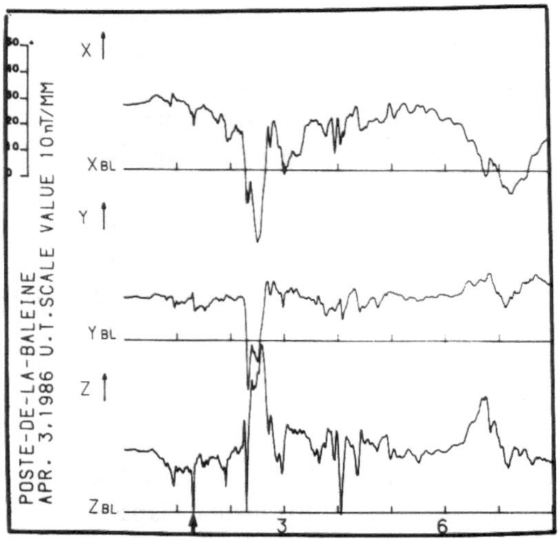

Fig. 2. A normal magnetogram obtained at a typical standard observatory (in this case Post-de-la-Baleine in Northern Canada). Until the early 1970's, such records alone were commonly used to diagnose the activity level of the magnetosphere. The perturbation identified by the vertical arrow at ~115 UT is associated with a substorm intensification (whose auroral signature is seen in Figure 1) extending over more than one time zone azimuthally. The large event seen in the following hour represents the sum of the effects of many such localized intensifications occurring one after the other.

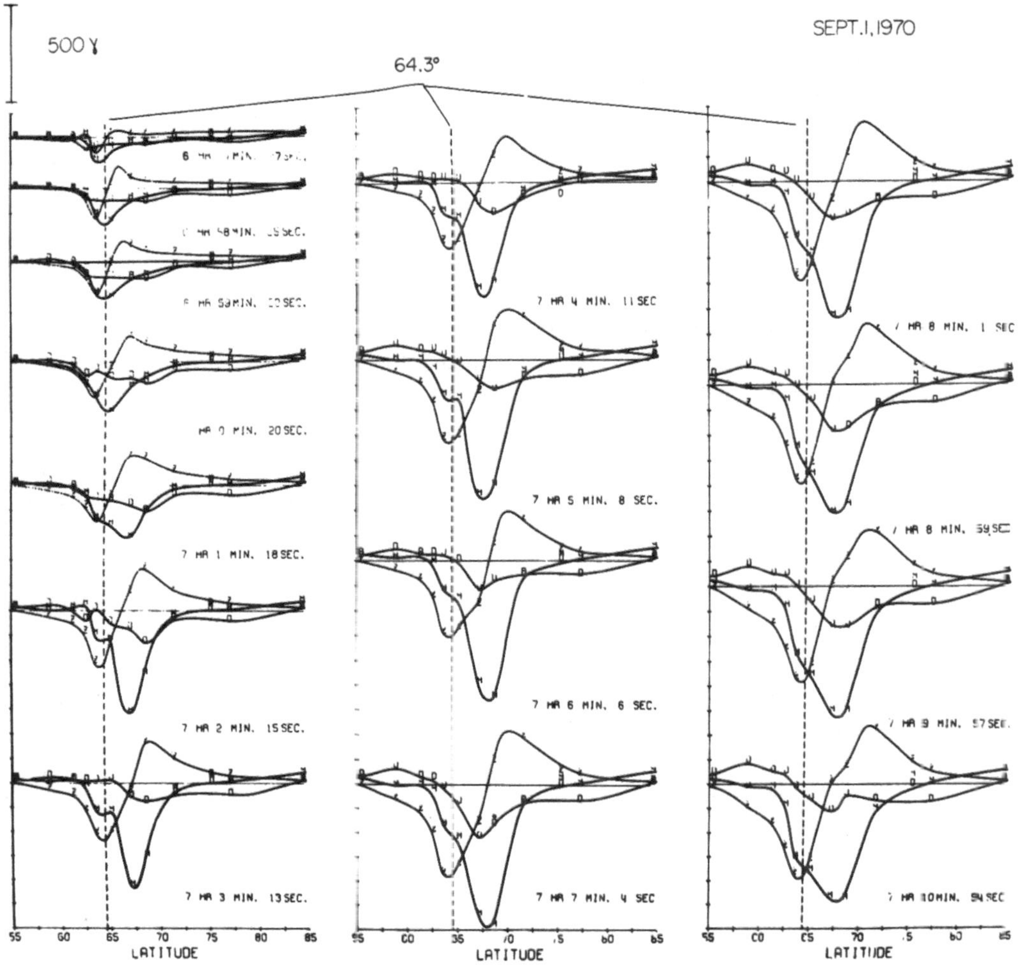

Fig. 3. Sequence of latitude profiles taken at approximately one minute intervals during the evolution of a substorm expansive phase which took place on September 1, 1970. The vertical dashed line marks the latitude of the initial activation and each data point represents an observation at an observatory on the University of Alberta meridian line of magnetometers. The meridian line configuration permitted the detailed study of the evolution of the latitudinal structure of the disturbed portion of the auroral oval in terms of ionospheric and field-aligned currents (after Kisabeth and Rostoker, 1974).

grams. For example, the sharp short-lived perturbation around 0115 UT (marked by the vertical arrow) could be interpreted as the signature of a localized westward electrojet whose center is to the north of the observatory. The brief positive D-component excursion followed by a longer negative perturbation is recognized as the edge effect of an auroral surge. (This surge appears in the center of the field of view in Figure 1.) However, a single observatory is very limited in its capability to define the latitudinal structure of the auroral electrojets. Figure 3 shows a sequence of latitude profiles taken at one minute intervals during the development of a substorm expansive phase by a meridian line of magnetometers across the auroral electrojet. This mode of data presentation shows the stepwise development of the substorm westward electrojet as it expands poleward during the expansive phase, with the latitudinal structure of the evolving electrojet being clearly apparent. One can imagine how an array of magnetometers along a line of constant longitude would be equally capable of detecting auroral structure in the east-west direction (cf. for example, Tighe and Rostoker (1981)).

While analysis of ground magnetometer data led to a reasonable forward model of the magnetosphere-ionosphere current systems associated with magnetospheric substorms (e.g. Hughes and Rostoker, 1979), a more sophisticated approach to the problem using large computers has emerged in recent times. This new approach featured the construction of an equivalent current system using the horizontal magnetic perturbation vectors from a reasonably dense and well distributed network of ground stations. The essence of the technique was developed in the 1970's independently by V. Mishin and co-workers in the U.S.S.R. and Y. Kamide and co-workers in the U.S.A. (cf. Bazarzhapov et al., 1979; Mishin et al., 1980; Kamide and Matsushita, 1979; Kamide et al., 1981). In briefly describing the technique, I shall follow the approach of Kamide et al. (1981) who constructed, for each instant in time, an equivalent current system whose flow lines were described by the scalar potential function Ψ such that the equivalent current density (in A/m) is given by

$$\mathbf{j}_T = \mathbf{n}_r \times \text{grad } \Psi \qquad [1]$$

where $\mathbf{n}_r$ is a unit radial vector. The total height integrated current was then represented as the sum of the equivalent current and a potential current $\mathbf{j}_P$, viz.

$$\mathbf{j} = \mathbf{j}_T + \mathbf{j}_P \qquad [2]$$

where div $\mathbf{j}_T = 0$ and curl $\mathbf{j}_P = 0$. This leads to the field-aligned current density being given by

$$j_{\parallel} = \text{div } \mathbf{j} = \text{div } \mathbf{j}_P \qquad [3]$$

where $j_{\parallel}$ is positive downwards.

The horizontal real ionospheric current is related to the electric field **E** (in a frame of reference corotating with the Earth) by

$$\mathbf{j} = \Sigma_P \mathbf{E} + \Sigma_H \mathbf{E} \times \mathbf{n}_r \qquad [4]$$

where Σ_P and Σ_H are respectively the height-integrated Pedersen and Hall conductivities. Considering the electric field

$$\mathbf{E} = -\text{ grad } \Phi \qquad [5]$$

to be electrostatic and perpendicular to the magnetic lines of force (i.e. magnetic field lines are equipotentials), one can obtain a partial differential equation for Φ in terms of Ψ by equating the right hand sides of Equations (2) and (4) and taking the curl of the resulting expression. In spherical coordinates, where θ is the colatitude and λ is the east longitude, the partial differential equation is written as

$$A\frac{\partial^2\Phi}{\partial\theta^2} + B\frac{\partial\Phi}{\partial\theta} + C\frac{\partial^2\Phi}{\partial\lambda^2} + D\frac{\partial\Phi}{\partial\lambda} = F \qquad [6]$$

where the coefficients (A,B,C,D,F) are complicated functions of (θ, λ, Σ_H, Σ_P, Ψ). Equation (6) can be solved using a finite difference scheme and reasonable boundary conditions for Φ at the pole and the equator. Once Φ is obtained, it is possible to calculate **E**, **j**, and $j_{\parallel}$ at all points in the ionosphere if one assumes some sort of reasonable ionospheric conductivity distribution. Figures 4a-d show the calculated electric potential and current distributions in the high latitude ionosphere for averaged activity levels in May and June of 1965. Clearly this technique represents a powerful extension of the simple magnetogram analysis methods of the past. It is not, however, without problems. In the first place, the real conductivity distribution in the ionosphere cannot easily be represented by any simple analytic formalism (as was used by Kamide et al. in their computations), a fact which is clearly evident on inspection of Figure 1. It may well be necessary to infer the high latitude conductivity distribution from some satellite borne detector (such as the UV imager flown aboard the Viking satellite) rather than rely on

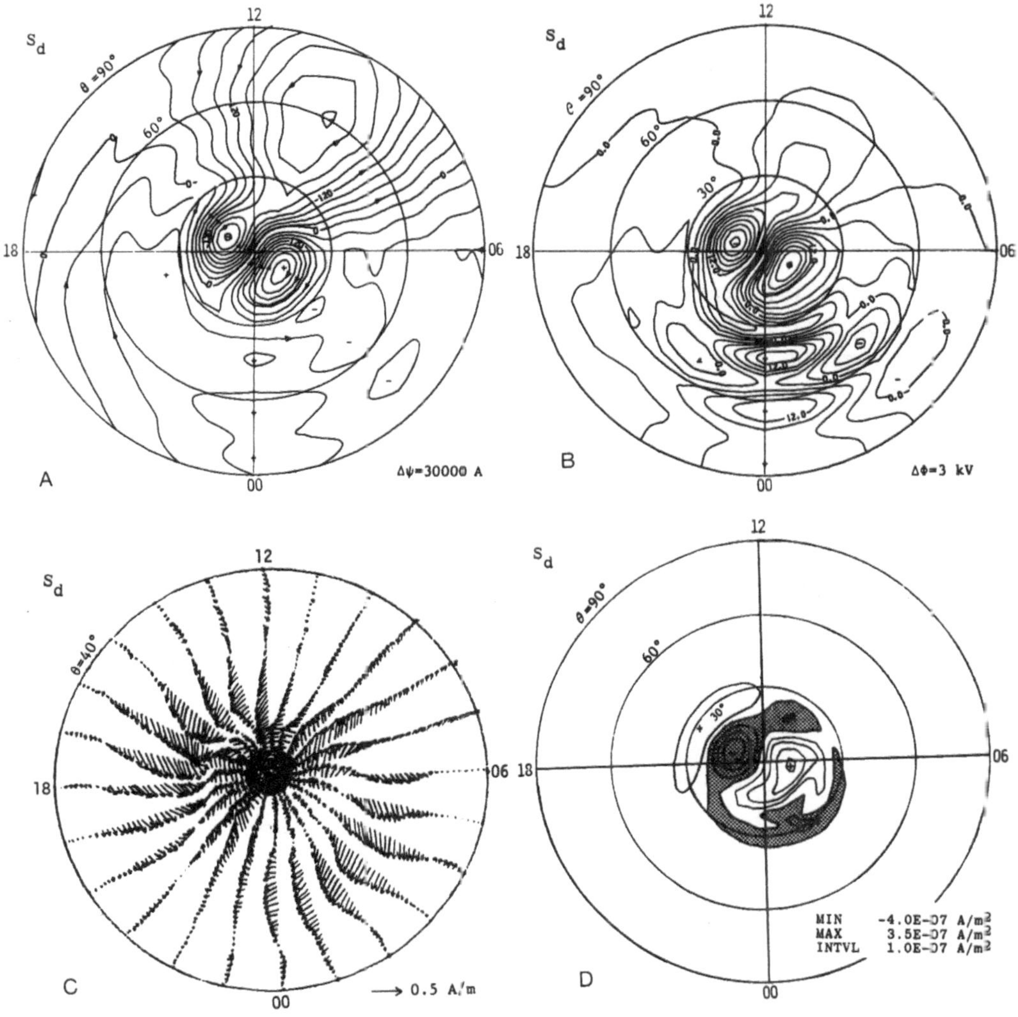

Fig. 4a. Equivalent current system for disturbed days calculated by averaging data over the period May-June 1965. Such plots are constructed using ground magnetometer data from a global array of stations and represent the input data for the KRM algorithm (after Kamide et al., 1981).
Fig. 4b. Equipotential contours constructed using the KRM algorithm for the input data presented in Figure 4a (after Kamide et al., 1981). From diagrams such as these, it is possible to estimate the cross polar cap potential drop.
Fig. 4c. Real ionospheric current vectors calculated using the KRM algorithm for the input data presented in Figure 4a (after Kamide et al., 1981).
Fig. 4d. Field-aligned current distribution calculated using the KRM algorithm for the input data presented in Figure 4a (after Kamide et al., 1981). The shaded region indicates upward field-aligned current flow.

some averaged distribution inferred from ground based remote sensing devices. More important is the fact that the distribution of parameters shown in Figure 4 is not unique. This is because the equivalent current $\mathbf{j}_T$ is not strictly toroidal since some part of the magnetic perturbation which was used to compute the equivalent current system is, in fact, due to the effects of non-solenoidal field-aligned currents. While it is possible to minimize the effects of these nonionospheric current contributions by assuming a reasonable forward model (eg. that of Hughes and Rostoker (1979)), the non-unique character of the computed parameter distributions should not be forgotten. Nonetheless, the application of the KRM algorithm discussed above can be seen to provide to the satellite experimenter an excellent tool for establishing the global state of the auroral ionosphere.

2.2. *Riometers*

Riometers are basically receivers of radio waves in the frequency range of 30-50 MHz which sample cosmic radio noise radiated by stellar sources in the galaxy. If there is energetic electron precipitation in the ionosphere over the receiving (broadbeam) antenna, the ionization generated by collisions of the primary particles with the atmospheric constituent particles shields the antenna on the ground from the incident cosmic noise. The signature of overhead energetic electron precipitation is, therefore, a decrease in cosmic noise level as shown in Figure 5. Until recently, riometers normally could only signal the presence of precipitation directly overhead. However, in recent times T.J. Rosenberg and colleagues have developed an imaging riometer system (IRIS) which can produce a full sky image of cosmic noise absorption at intervals of as little as one second (cf. Detrick and Rosenberg, 1988). The IRIS antenna is a square array of 64 circularly polarized crossed dipole antennas oriented parallel to the Earth's surface and located a quarter wave length above a conducting ground plane. The signals from

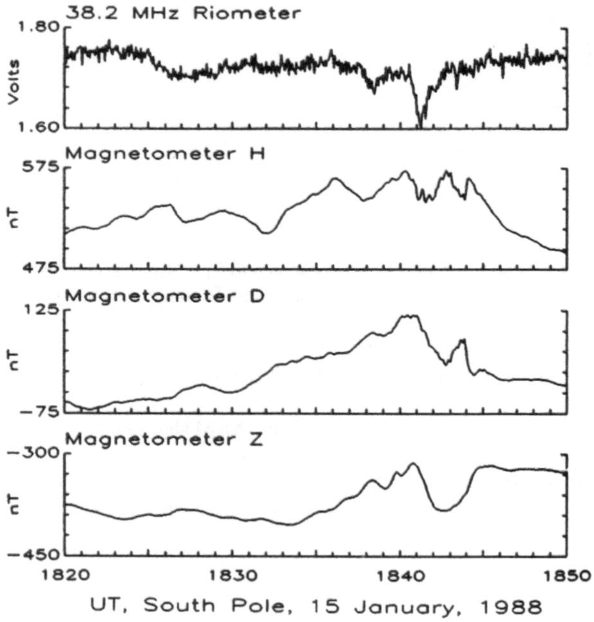

Fig. 5. Three components of the perturbation magnetic field and riometer measurements recorded at South Pole station on January 15, 1988 (courtesy T.J. Rosenberg). By a suitable conversion parameter, the reading in volts can be converted to a measurement in dB and the final data would be presented with quiet time variations extracted. The decreases in cosmic noise level indicate enhanced conductivity over the instrument site due to energetic particle precipitation.

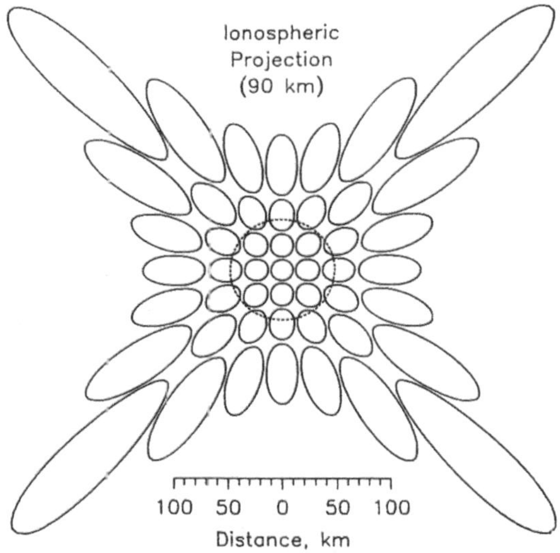

Fig. 6. Projection of each IRIS beam (-3 dB contours) onto the ionosphere at an altitude of 90 km. The dashed circle in the center of the field of view represent the area sampled by a conventional broad beam riometer (after Detrick and Rosenberg, 1988).

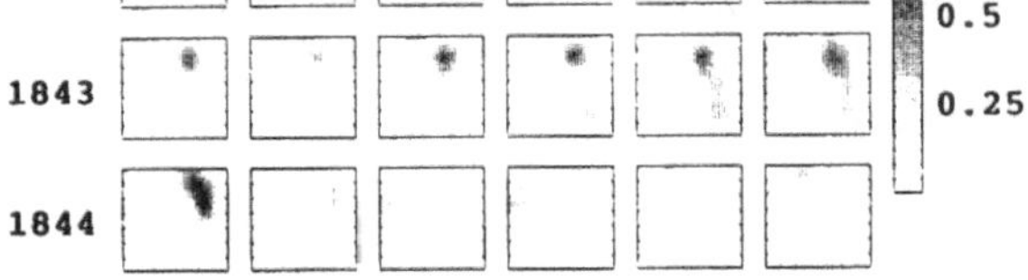

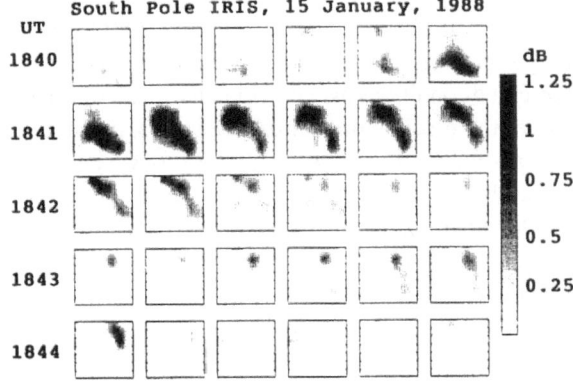

Fig. 7. Sequence of images taken at 10 s intervals by the IRIS system showing the evolution of cosmic noise absorption over the instrument site at South Pole station for part of the event shown in Figure 6 (after Detrick and Rosenberg, 1988).

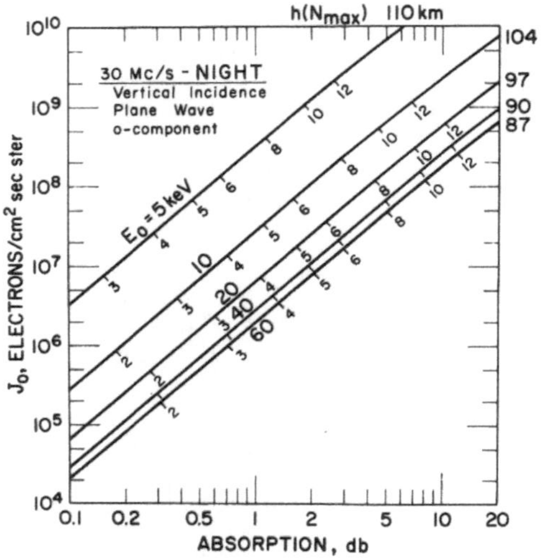

Fig. 8. Integrated absorption of cosmic noise due to energetic electron precipitation characterized by flux J_o and energy E_o (after Bailey, 1968). Note that the same amount of absorption can be generated by a small flux of 60 keV electrons or a larger flux of 10 keV electrons.

the dipoles are combined in a complex of 15 Butler matrices which provide the phasing needed to produce 49 independent beams in a 7 x 7 square array as shown in Figure 6. An example of a transient absorption event is shown in Figure 7, from which it is evident that this instrument is able to monitor a considerable portion of the sky overhead in terms of quantifying the level of auroral electron precipitation and identifying spatial and temporal variations in that precipitation. It is readily apparent that this information should be of great value to any investigator with satellite data obtained on field lines penetrating the ionosphere in the vicinity of the riometer. This ground technique has the obvious advantage over optical techniques in that it is not dependent on cloud conditions over the riometer site. It would be nice to be able to define the energy spectrum of the precipitating auroral electrons, however Figure 8 shows that this may prove a difficult task indeed. Evidently small fluxes of very energetic electrons are capable of producing the same absorption levels as larger fluxes of less energetic electrons. Nonetheless, it is quite clear that imaging riometer systems have an important role to play in defining regimes of auroral particle precipitation which cannot help but be of use to researchers trying to interpret their satellite data.

2.3. Auroral Radar Systems

As pointed out earlier in this paper, it is imperative to gain some measure of knowledge of the large-scale electric field distribution in the magnetosphere-ionosphere system if one is to ever hope to understand the nature of the solar-terrestrial interaction. Fortunately, auroral radars are able to obtain this information by bouncing high frequency radio waves off irregularities in the auroral ionosphere and inferring the line-of-sight flow velocity of the local plasma from the Doppler shift of the return signal picked up by appropriately placed ground receivers. There are two basic types of auroral radar techniques which I shall describe below.

Coherent scatter. Coherent scatter radars detect backscatter from density irregularities associated with plasma waves in the E- and F-regions of the ionosphere. The irregularities are thought to be excited through the presence of an ambient electric field and, unless that electric field exceeds 15 mV/m the lack of irregularities makes it impossible to measure that electric field which may be present (at least in the E-region). In order to get enough backscattered power, scattering must occur off several irregularities whose periodic spacing must be such that the Bragg condition is met leading to the selective observation of irregularities with appropriate wave numbers. The Doppler shift in the backscattered signal is considered to be a reasonable indicator of the component of the particle drift velocity along the line of sight of the radar. While this seems to work well for drift velocities < 700 m/s in the E-region of the ionosphere, it appears that there is a saturation effect above that critical velocity where the Doppler shift from the irregularities no longer corresponds to the drift velocity of the particles. This problem does not seem to arise for coherent backscatter from F-region irregularities.

As mentioned earlier, a single radar can only measure the Doppler shift associated with particle drifts along the line of sight. However, it is possible to use two radars with transmitters well separated so that Doppler shifts can be measured for particle drifts along two different lines

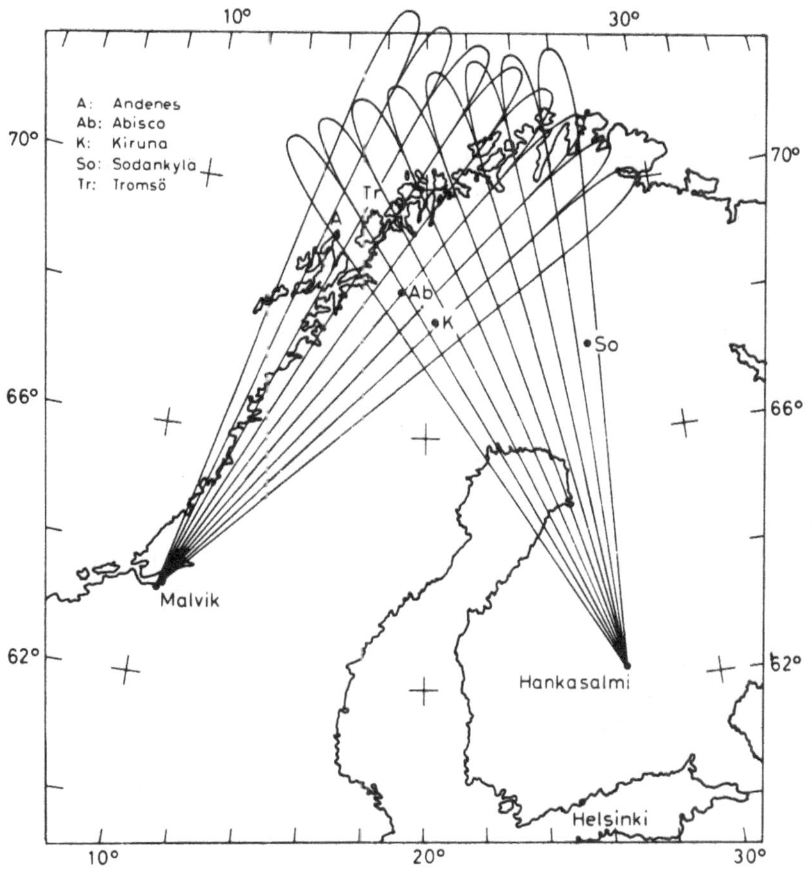

Fig. 9. Beam patterns of the two transmitters of the STARE radar which has operated in Scandinavia over the past decade (after Walker et al., 1979). The electric field vectors are measured from the Doppler shift of plasma irregularities in the region where the beams intersect.

of sight. This permits the construction of the total drift velocity at the intersection points of the two transmitter beams and hence the measurement of the total vector ionospheric electric field. The first real exploitation of this technique involved the deployment of the Scandinavian Twin Auroral Radar Experiment (STARE) by R.A. Greenwald and colleagues (cf. Greenwald et al., 1978). With one transmitter at Malvik (Norway) operating at 140 MHz and a second transmitter at Hankasalmi (Finland) operating at 143.8 MHz, a significant area of the ionosphere over northern Norway could be sampled (Figure 9) at intervals as short as 20 s. It was therefore possible to create two-dimensional plots of the electric field vectors which were of potentially great value to satellite experimenters (Figure 10). While the STARE radar and radars like it scattering off of E-region irregularities suffer from saturation effects during episodes of particularly high electric field, lower frequency HF radars set up recently to scatter off of F-region irregularities (cf. Greenwald et al, 1985) look very promising in providing additional sources of information on ionospheric electric fields. At present, two of these HF radars are operating in eastern Canada with a third operating near the conjugate point in Antarctica.

Incoherent Scatter. Incoherent scatter radars detect backscatter off thermal density fluctuations in the ionospheric plasma with scale sizes smaller than the wavelengths of the radio waves. Each irregularity scatters independently of the others so the signal returned to the ground receiver contains a mixture of phases, viz. it is incoherent. Like its coherent scatter counterpart, the backscattered waves contain information about the convection electric field normal to the line of sight of the radar. However, the intensity of the backscattered signal can be used to evaluate the electron number density in the height range being sampled. In addition, thanks to a type of coherent scatter which takes place simultaneously, one can infer the ion temperature and the ratio of electron temperature to ion temperature. In the tristatic configuration adopted for the European Incoherent Scatter (EISCAT) facility, it is possible to measure both the component of the electric field normal to the magnetic field and the flow of plasma along the magnetic lines of force. This latter capability is of particular value in the light of the importance being attached nowadays to the ionosphere as a source of plasma for the plasma sheet.

It is worth pointing out that incoherent scatter radar can be a particularly powerful tool in monitoring the position of the convection reversal boundary normally located in the poleward portion of the auroral oval. Figure 11 shows a plot of the borders of the auroral electrojets as a function of Universal Time for a period of time during which a magnetospheric substorm (onset 2131 UT) took place. During the latter stages of this interval (~2228 UT onwards) the imager aboard the Viking satellite was able to track the development of the substorm. However, despite the fact that the imager can define the luminosity patterns, it cannot establish the position of the convection reversal. Fortunately, for this event, a chain of magnetometers along the west coast of Greenland was able to define the borders of the eastward and westward electrojets and the position of the convection reversal boundary (i.e. the Harang discontinuity) in the evening sector. However, the Sondrestrømfjord incoherent scatter radar was also operating at this time and was also able to define the convection reversal boundary as well as providing crucial information on electric field magnitude and direction across the electrojets and the electron number density over the region being scanned. It is clear that this information was/is an important complement to the luminosity data acquired by the Viking imager in terms of tracking the development of the magnetospheric substorm under study.

2.4. *Ground Based Detectors of Auroral Luminosity*

It has been a time honored tradition in space science to study the fluctuations of the aurora in the night sky using first the human eye and then moving on to detectors which permitted the

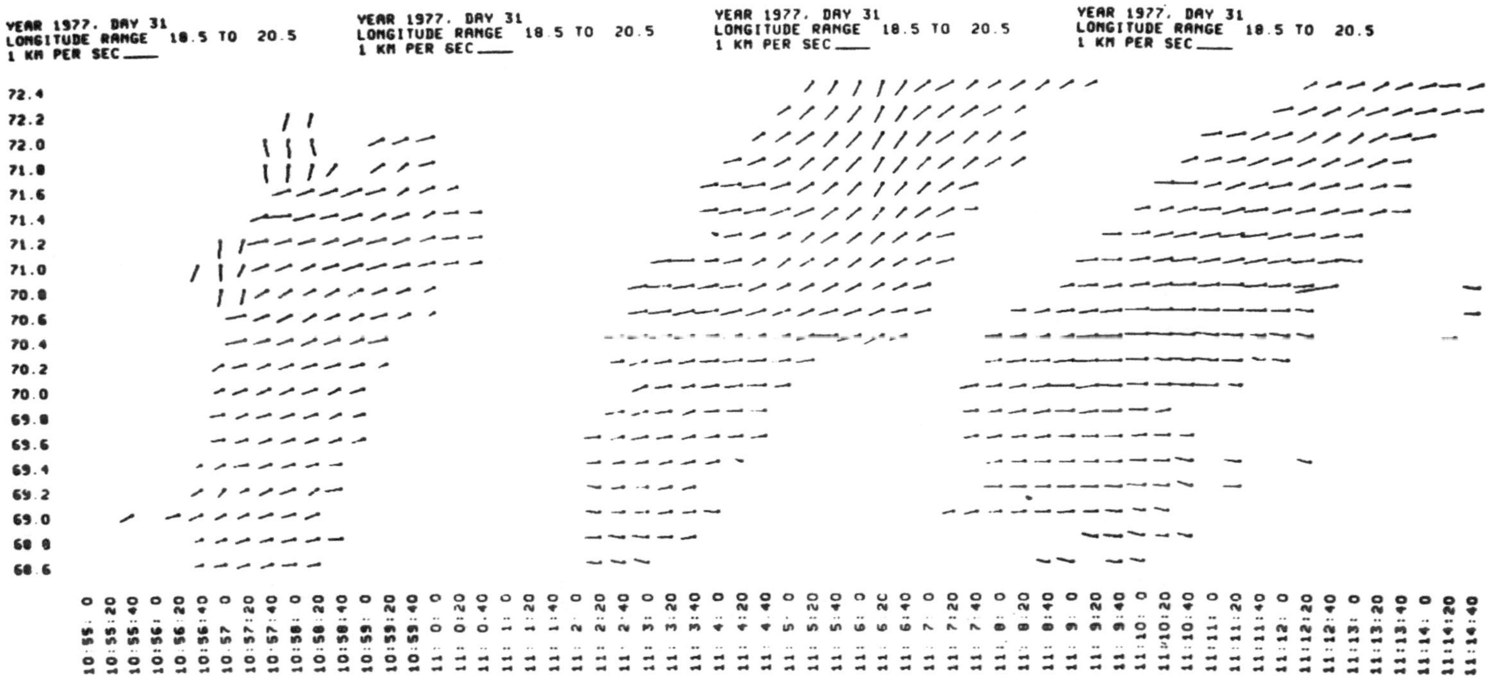

Fig. 10. Mean irregularity velocity inferred from scattered power integrated over 20 s intervals for the period 1055:00-1114:40 UT on January 31, 1977 (after Walker et al., 1979). Blank areas indicate the absence of irregularities from which the transmitter signals can be scattered and hence a low electric field. The electric field can be inferred from such data as assuming the irregularities drift at the E x B convection velocity. Long period pulsations in the inferred electric field are evident in these data.

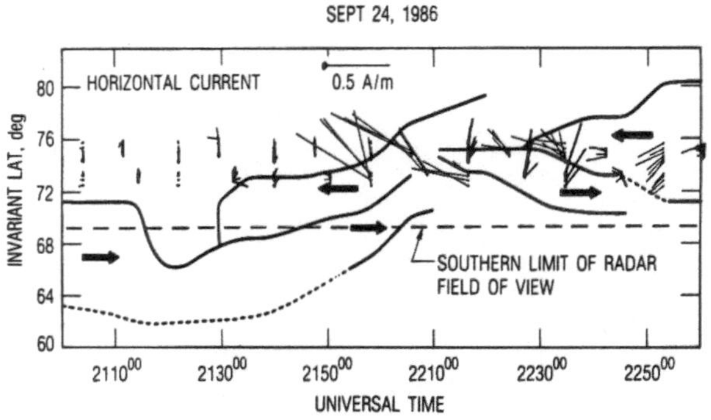

Fig. 11. Horizontal ionospheric current vectors inferred from data obtained by the Sondrestrømfjord incoherent scatter radar during a substorm event on September 24, 1986. The ability of the radar to obtain information about both electric field and ionospheric number density permits the raw ion drift data to be converted to information about electric current flow. The current vectors are superposed on contours of the poleward and equatorward borders of the eastward and westward electrojets as inferred from a meridian line of magnetometers along the western coast of Greenland. Both the magnetometer data and the radar data can identify the location of the convection reversal boundary (viz. the Harang discontinuity in the evening sector) which is of crucial importance in determining the configuration of the magnetosphere during any individual event (after Lyons et al., 1989).

information on the auroral light to be documented in a quantitative fashion. The first detectors were simple spectrographs (cf. Størmer, 1955) which permitted the primary spectral lines of the aurora to be identified. Later on the rich complexity of auroral spectra was charted using spectrophotometers scanning over limited wavelength ranges (cf. Vallance Jones, 1964). However, it was not until the development of the all-sky camera that the phenomenology of auroras as a signature of the solar terrestrial interaction began to emerge (Feldstein, 1962; Akasofu, 1964). At present, the following methods are available to monitor auroral luminosity from the Earth's surface:

All-sky Camera. Early versions of this instrument involved a configuration of mirrors which allowed the whole sky to be photographed automatically at preset intervals and exposure times. In the days of low ASA film, the exposure times were of the order of many seconds and it was virtually impossible to identify diffuse auroras in the photographs. Thus auroral dynamics were judged purely on the basis of the evolution of bright discrete arcs. In addition, the nature of the optics of the all-sky camera dictated that only the overhead arcs appeared in a non-distorted fashion with strong distortion being evident at the edges of the image. Some of this distortion could be removed by linear mapping techniques, however, interpretation of the auroral forms more than 200 km away from the zenith position was always somewhat suspect. In recent times, the speed of the film has permitted the exposure times to drop to 1 second and the use of colour film has increased the amount of information that can be obtained on the character of the auroral luminosity. However, there are few all-sky cameras operating in the world today with the Finnish group of R. Pellinen and co-workers being the foremost exponents of this method of auroral imaging nowadays.

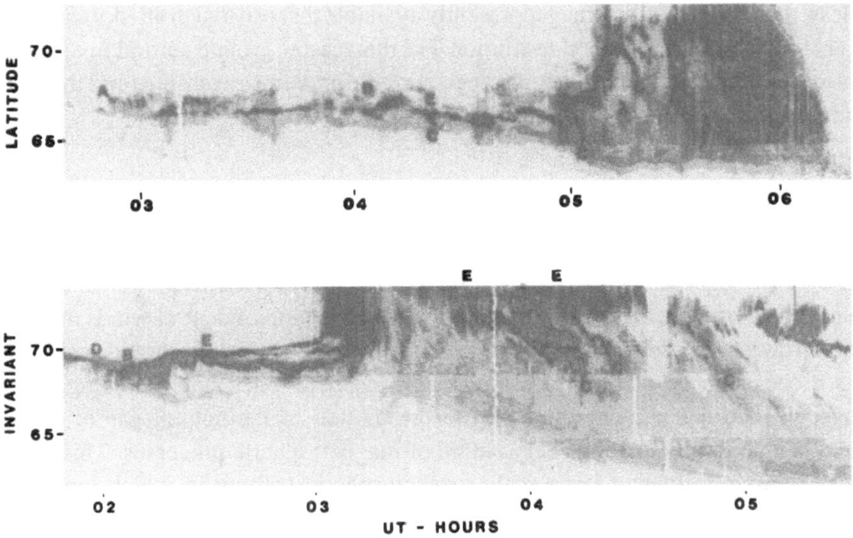

Fig. 12. Merged data from three meridian-scanning photometers operating in central Canada during the interval 0245-0615 UT on March 3, 1978 (top panel) and 0150-0530 UT on February 5, 1978 (bottom panel). The poleward expansion of the disturbed region and equatorward drifting forms are clearly evident using this format (after Atkinson et al., 1989).

Meridian Scanning Photometer. Hunten (1955) introduced the scanning photoelectric photometer which allowed the auroral intensity to be measured at selected wavelengths (determined by filters) along a north-south strip across the night sky. Typical scan times from horizon to horizon are of the order of 30 s and sequential scans produce a record of the latitudinal distribution of auroral luminosity as a function of time along the meridian of observation (as shown in Figure 12). Records of this type have been used to great effect by A. Vallance Jones and co-workers in Canada (e.g. Creutzberg et al., 1981) and by R.H. Eather and colleagues (e.g. Eather, 1984) operating at the South Pole station to follow poleward and equatorward development of discrete auroral arc features during magnetospheric substorm activity studying the ratios of the intensities of the various spectral lines, it is possible to gain a measure of information about the fluxes of the energetic electrons and protons at different energies precipitating into the auroral ionosphere.

Television Camera. With the development of low light level television monitors, the door was opened to vastly improved capabilities for the study of auroral structure and fast time variations. From a situation where images were recorded at intervals of 1 s or more, suddenly a timing accuracy of the order of 50 ms became possible. This kind of timing accuracy is essential in the correlation of fluctuations in auroral luminosity with the bursts of high frequency noise found in the auroral oval during such phenomena as pulsating aurora (cf. Tsuruda et al., 1981). For those space scientists interested in plasma instabilities occurring in the ionosphere which have manifestations in the precipitation of auroral particles, the development of low light level TV imagers has provided for a set of observations which can help to distinguish amongst the various instability mechanisms which may be operative in the auroral oval.

It goes without saying that any ground observations of auroral luminosity are restricted by the need for a clear sky over the imaging device. This is not a trivial constraint if one is operating arrays of detectors in an attempt to characterize the global nature of auroral activity.

Nonetheless, there is no satellite imager presently available that can match all-sky cameras and TV imagers for spatial and temporal resolution. For this reason, ground auroral imagers are an important tool to the satellite investigator who requires information regarding fine structure and rapid temporal variations of the aurora.

2.5. *Ground Based Detection of Neutral Winds*

For over two decades, it has been well recognized that **E** x **B** drift motion of ions in the F-region of the ionosphere and below can lead to the production of neutral winds through the process of ion drag (cf. Fedder and Banks, 1972). However, neutral winds should not be viewed simply as a consequence of the process of magnetosphere-ionosphere coupling. There is increasing evidence that the energy stored in the motion of the neutral gas of the upper atmosphere is able to influence the very nature of magnetospheric processes. This might be expected in periods following large scale magnetospheric activity in which ion drag has produced significant neutral winds in and below the F-region. When the convection electric field has died down, collisions of the neutrals with the ionospheric ions can lead to polarization electric fields and ionospheric current flow (eg. Rostoker and Hron, 1975; Lyons and Walterscheid, 1985). These feedback effects are so important in the context of solving the magnetosphere-ionosphere coupling problem that it is vital to complement ionospheric particle and field measurements with comprehensive measurements of the thermospheric neutral wind patterns.

Perhaps the most promising method of remote sensing thermospheric neutral winds utilizes Fabry-Perot interferometers to measure the line-of-sight Doppler shift of radiation from neutral atoms. In principle, three properly placed interferometers (viz. a tristatic array) can be used to evaluate the horizontal and vertical components of the neutral wind flows within the common field of view of the instruments. Rees et al. (1984) have developed a Doppler imaging system (DIS) which is based on a field-widened Fabry-Perot interferometer that simultaneously records a number of Fabry-Perot fringes using a very stable, high resolution and high speed imaging photon detector. Figure 13 shows the configuration of the system along with the specifications of the instrument components. This particular system records the OI 630 nm emissions of atomic oxygen and produces velocity vectors as shown in Figure 14. In the light of the fact that neutral winds are of consequence both in the estimation of ionospheric conductivity (Akasofu and Dewitt, 1965) and for feedback effects in the magnetosphere-ionosphere coupling process, it is clear that ground based instrumentation of the type described above is crucial if one is to be able to fully describe in a quantitative fashion the nature of the solar-terrestrial interaction.

In this section, I have outlined the various ground based remote sensing techniques which are available to the space researcher who wishes to know, at some instant in time, the characteristics of the disturbed auroral ionosphere. If one wishes to answer the question as to why such remote sensing methods still have importance in an era when some *in situ* measurements can be made, one needs only to recognize the global nature of the magnetosphere-ionosphere coupling process. It is quite unacceptable for a satellite researcher to be unconcerned about the precise nature of the activity at the foot of the magnetic field lines on which the satellite measurements are being made. The presence of sharp discontinuities in auroral luminosity and electric current flow in the auroral ionosphere are testament to the fact that satellites might expect to encounter similar sudden spatial changes in particle and field properties which might easily be mistaken for temporal variations. There is little doubt about the continuing importance of ground based remote sensing in the future of space research.

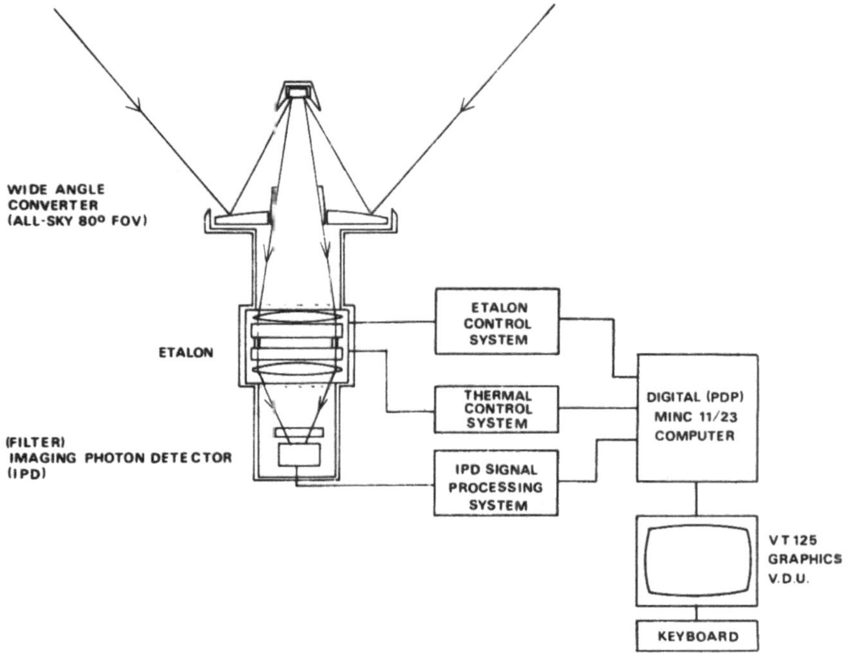

A. INTERFEROMETER
Fabry–Perot etalon
Diameter 132 mm
Plate separation 14 mm
Plate defect (after coating)
$<\lambda/280$ (at 546 nm)
Combined etalon defects
$<\lambda/150$ (at 546 nm)
[Plate, coating, parallelism etc.]
Thermal expansion coefficient of spacer (cemented)
4×10^{-8}/°C
Pre-filter 1 nm FWHM at 630.5 nm

B. WIDE ANGLE CONVERTER
Spherical primary mirror
Spherical secondary mirror
Tertiary (lens) 135 mm diameter
635 mm EFL
Matching 80° full angle FOV onto 6 fringes of the 14 mm gap
Fabry–Perot etalons
[Operates as a teleconcentric converter]

C. IMAGING SYSTEM
[Imaging Fabry–Perot fringes onto IPD]
~510 mm equivalent focal length
135 mm diameter

D. IMAGING PHOTON DETECTOR
18 mm photocathode (S25)
4 microchannel plate intensifier
Low resistance resistive anode position encoder
512×512 pixels FWHM
150 kHz max. photon counting rate

E. COMPUTER
PDP 11/23 (MINC)
256 K bytes
Using RT11/XM and RX02 drives with DMA
(DRV 11 B) interface

Fig. 13. Instrument configuration and hardware specifications of a computer controlled Doppler imaging system developed by D. Rees and colleagues at the University College London (after Rees et al., 1984). Using two instruments in a bistatic mode or three instruments in a tri-static mode can allow the horizontal or total velocity vectors respectively to be evaluated in principle.

3. THE IMPORTANCE OF SOUNDING ROCKET MEASUREMENTS

There is a long and time honoured history of measurements of the properties of the auroral ionosphere using detectors flown aboard sounding rockets. It is quite reasonable to ask, however, whether sounding rockets are an efficient vehicle on which to place particle and field detectors in the light of the fact that a satellite with appropriate orbital elements seems quite capable of sampling the same region of space. Since the rocket payload samples the designated region of space only once before the experiment is concluded, it would seem a grossly

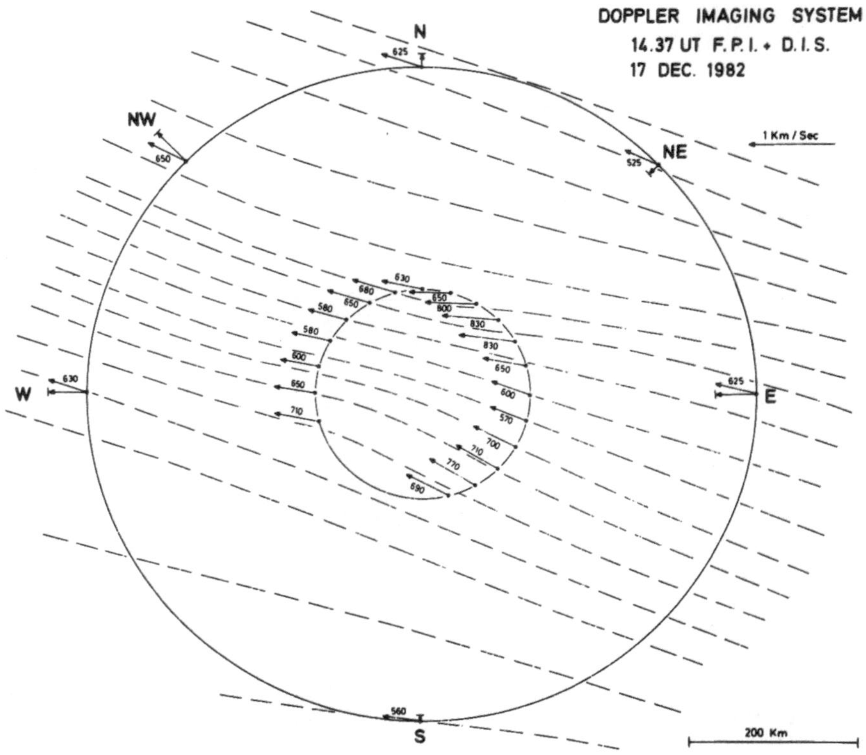

Fig. 14. Type of output expected from a Doppler imaging system in terms of neutral wind velocity vectors (after Rees et al., 1984). It is clear that neutral winds can acquire velocities of hundreds of m/s and the feedback effect on the magnetosphere of this storage of energy and momentum in the neutral atmosphere cannot be ignored.

inefficient method of studying the ionosphere compared to the launch of a satellite whose detectors could sample the same region of space any number of times. In fact, one's first tendency is to view a rocket as a satellite with a short life time and strange orbital elements. On closer inspection, this turns out to be an unfair appraisal of the benefits stemming from sounding rocket experiments.

In the first place, if one is interested in the microphysics of auroral processes, the altitude range sampled by rockets (viz. a few tens of km to over a thousand km) can be viewed as a giant plasma laboratory with no walls. There have been several important studies in the recent past where sounding rocket payloads have been used to study such crucial physical problems as auroral particle acceleration (cf. Bryant, 1981) and plasma turbulence (cf. Kelley et al., 1982). Of course, it is not unreasonable to ask whether or not the same payloads carried by the sounding rockets could not have been launched on polar orbiting satellites and thus considerably more data taken over a lengthier period of time. While the instruments could well have been flown aboard a satellite, there is one aspect of a rocket experiment which cannot be reproduced using orbiting satellites. When a rocket is near apogee, it is practically stationary with respect to the background plasma. This allows a given region of space to be sampled for a significant period of time compared to normal plasma time scales. In contrast, a polar orbiting satellite passes through a similar volume of space very quickly due to its orbital velocity (typically of the order of several km/s). It is very difficult to study plasma processes in an inherently inhomogeneous medium when spatial and temporal changes are mixed and the detector spends little time in the volume of plasma being sampled.

The second important use of sounding rockets is as a vehicle to transport material which is released in the ionosphere in an effort to monitor plasma and field parameters over an extended region of space or in an attempt to alter processes which are already taking place. In the first category, the most noticeable example is the launch of shaped barium charges which are detonated in such a way as to jet barium ions along the magnetic lines of force. The moving barium cloud is able to sense the presence of parallel electric fields by exhibiting differential drift velocities perpendicular to the magnetic lines of force as a function of altitude (cf. Wescott et al., 1976). An example of the second category of active experiment is the injection of a cloud of water vapor into a region of the ionosphere where active auroral arcs are present in order to see if the auroral activation process is altered (cf. the "Waterhole" project described in Whalen et al. (1981)). The ability to inject material at a specified place and at a time decided by the experimenter reflects the effectiveness of sounding rockets in the role of facilitating active experiments.

Finally, although this does not relate directly to the science that can be carried out, it is worth pointing out that in this day of lengthy delays between the conception of a satellite experiment, its construction, launch and the ultimate analysis of the data, sounding rockets provide the possibility of a rapid movement from concept to reality. For the purpose of training a modern cadre of space scientists (particularly experimentalists), the possibility of carrying out a space experiment in a time scale consistent with graduate study programs is attractive indeed. Sounding rockets may yet prove to be a vehicle through which space science can rejuvenate itself while allowing specific problems to be tackled which can be carried out quickly and for a relatively low cost compared to typical satellite experiments.

4. THE IMPORTANCE OF MEASUREMENTS USING BALLOON BORNE DETECTORS

Balloon borne detectors have provided space scientists with clues to the nature of our charged particle environment long before rockets and satellites were able to make the *in situ* observations which gave us the direct measurements necessary for any quantitative studies of geospace. The early detectors were designed to investigate cosmic rays, however it was not long before it became evident that it was possible to measure bremsstrahlung X-rays due to collisions between fast auroral electrons and the atoms and molecules of the neutral atmosphere (Winckler and Peterson, 1957). For the next two decades, there were a large number of studies of X-ray signatures of auroral phenomena using detectors on single balloons and in multiple balloon launch scenarios. These types of experiments are rarely carried out nowadays because the primary electron fluxes can be measured directly by satellite and rocket borne detectors. However, in the late 1960's, it was discovered that magnetospheric electric fields could be measured at altitudes above about 10,000 m as long as the spatial scale size of the fields was greater than a few times the Earth-ionosphere separation (Mozer and Serlin, 1969). In the early years after this important realization, most experiments were designed to measure the magnetospheric electric fields and any contributions from atmospheric electric fields were considered to be a nuisance to be avoided. In recent times, however, it has become clear that much can be learned about both the atmospheric electric field and the electrical conductivity of the lower atmosphere (Holzworth, 1987). With the increasing recognition of the role played by the global electric circuit in coupling the lower regions of the "neutral" atmosphere to the ionosphere-magnetosphere system (Markson, 1979), it seems obvious that balloons represent

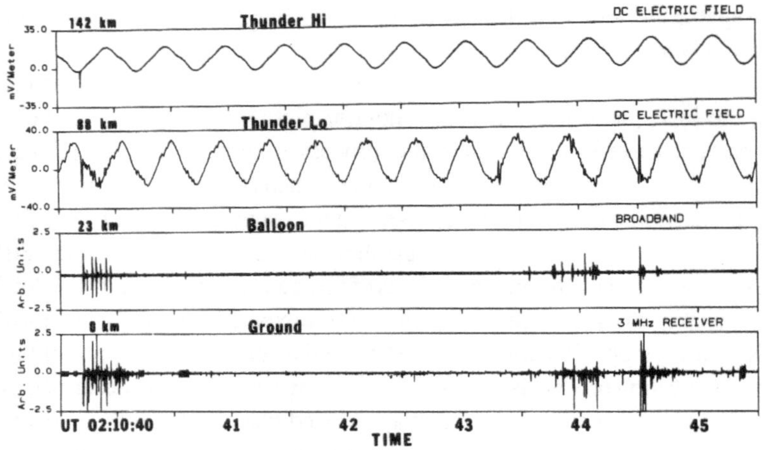

Fig. 15. Simultaneous data from electric field detectors aboard high altitude and lower altitude rockets together with VLF data from a detector aboard a balloon floating in the stratosphere below the rockets and a ground based 3-MHz receiver. The high frequency disturbances are due to lightning strokes. Such information is critical to improving our understanding of the global electric circuit (after Kelley et al., 1985).

a very useful vehicle for studying the lower atmosphere in an altitude range which does not lend itself to being easily probed using rocket-borne detectors. Figure 15 shows data from a campaign developed to study the thunderstorm electric field in the light of the importance which lightning plays in electrically coupling the lower atmosphere to the ionosphere. The electric field data in the figure show how balloons (as well as rockets) serve as useful vehicles which permit height profiles of key electrodynamic parameters to be obtained by monitoring that portion of the altitude range which cannot be probed by satellites making *in situ* measurements.

It is worth pointing out that the technology of ballooning has come a long way since the flights in the early days dedicated to cosmic ray studies. The new super-pressure balloons permit flight times in excess of a month while the RACOON type zero-pressure balloons have taken data for periods ranging from a few days to two weeks (although the altitudes of operation change by as much as 20 km daily as the balloon moves from daytime to nighttime conditions). Finally it should be noted that balloon borne detectors provide the cheapest method of obtaining continuous electric field data for periods up to a month or so. It is technically feasible to obtain data from detectors aboard several super-pressure balloons operating simultaneously using one launch site and satellite linked communications (R.H. Holzworth, personal communication). It is easy to see that balloons carrying electric and magnetic field detectors could provide important coverage of the electric and magnetic field pattern in regions where large auroral radars and magnetometer networks are not available (eg. over the oceans of the world). The additional magnetometer information could greatly improve the quality of the input data for algorithms such as that developed by Kamide et al. (1981) described earlier in this paper. In addition, the electric field measurements could be very useful in providing constraints for the algorithm by acting as check points for the predictions of the electric field configuration.

5. CONCLUSIONS

In this paper, I have tried to outline the modern ground based techniques for inferring information about the solar-terrestrial interaction and have shown that both sounding rockets and balloon borne measurement platforms still have an important role to play in modern space physics. In conclusion, I would like once again to emphasize that space physics suffers greatly from a paucity of data in terms of simultaneous measurements at different points in space. Ground based measurements yield the possibility of having, at all times, a reasonable evaluation of the electrodynamic and charged particle environment in the terrestrial ionosphere. When researchers finally develop a magnetic field model for the Earth which will permit mapping from the ionosphere into the outer magnetosphere, any measurement in the magnetosphere will be complemented by a second set of information at the foot of the field line on which the satellite measurement was made. Only with satellite and ground based remote sensing data complementing one another, will researchers have a potentially adequate data base which will permit them to have a realistic expectation of finding a solution to the problem of the solar-terrestrial interaction.

ACKNOWLEDGEMENTS

I am grateful to T.J. Rosenberg and R.H. Holzworth for information they provided to me regarding modern riometry and ballooning respectively. I also am indebted to H. Opgenoorth for useful discussions on the capabilities of incoherent scatter radar facilities. This research was supported by the Natural Sciences and Engineering Research Council of Canada.

REFERENCES

Akasofu, S.-I., 1964, The development of the auroral substorm, *Planet. Space Sci.*, 12:273.

Akasofu, S.-I., and DeWitt, R.N., 1965, Dynamo action in the ionosphere and motions of the magnetospheric plasma III. The Pedersen conductivity, generalized to take account of acceleration of the neutral gas, *Planet. Space Sci.*, 13:737.

Appleton, E.V. and Barnett, M.A.F., 1925, Local reflection of wireless waves from the upper atmosphere, *Nature*, 115:333.

Atkinson, G., Creutzberg, F., Gattinger, R.L., and Murphree, J.S., 1989, Interpretation of complicated discrete arc structure and behavior in terms of multiple X lines, *J. Geophys. Res.*, 94:5292.

Bailey, D.K., 1968, Some quantitative aspects of electron precipitation in and near the auroral zone, *Rev. Geophys.*, 6:289.

Barcus, J.R., and Brown, R.R., 1966, Energy spectrum for auroral-zone X-rays 2. Spectral variability and auroral absorption, *J. Geophys. Res.*, 71:825.

Bazarzhapov, A.D., Mishin, V.M., Shirapov, D.S., and Shpynev, G.B., 1979, Electric fields and currents in the quiet magnetosphere as determined from ground measurements, *Issle. Geomag. Aeron. Fiz. Solntsa,* 46:13.

Biermann, L., 1951, Kometenschweift und solare korpuskularstrahlung, *Z. Astrophys.*, 29:274.

Birkeland, K., 1908, The Norwegian Aurora Polaris Expedition 1902-1903, vol. 1, 1st sec., Aschhoug, Christiania.

Bryant, D.A., 1981, Rocket studies of particle structure associated with auroral arcs, *in* "Physics of Auroral Arc Formation", ed. by S.-I. Akasofu and J.R. Kan, p. 103, Geophysical Monograph 25, American Geophysical Union, Washington, D.C.

Carpenter, D.L., 1963, Whistler evidence of a "knee" in the magnetospheric ionization density profile, *J. Geophys. Res.*, 68:1675.

Chree, C., 1912, "Studies in Terrestrial Magnetism", MacMillan and Co. Ltd., London.

Creutzberg, F., Gattinger, R., Harris, F., and Vallance Jones, A., 1981, Pulsating auroras in relation to proton and electron auroras, *Can. J. Phys.*, 59:1124.

Detrick, D.L. and Rosenberg, T.J., 1988, IRIS: an imaging riometer for ionospheric studies, *Antarct. J. U.S.*, XXIII:196.

Eather, R.H., 1984, Dayside auroral dynamics, *J. Geophys. Res.*, 89:1695.

Fedder, J.A. and Banks, P.M., 1972, Convection electric fields and polar thermospheric winds, *J. Geophys. Res.*, 77:2328.

Feldstein, Y.I., 1963, Some problems concerning the morphology of auroras and magnetic disturbances at high latitudes, *Geomagnetizm i Aeronomiya,* 3:183.

Greenwald, R.A., Weiss, W., Nielsen, E., and Thomson, N.R., 1978, STARE: a new radar auroral backscatter experiment in northern Scandinavia, *Radio Sci.,* 13:1021.

Greenwald, R.A., Baker, K.B., Hutchins, R.A. and Hanuise, C., 1985, An HF phased-array radar for studying small-scale structure in the high-latitude ionosphere, *Radio Sci.,* 20:63.

Hau, L.-N., Wolf, R.A., Voigt, G.-H. and Wu, C.C., 1989, Steady state magnetic field configurations for the Earth's magnetotail, *J. Geophys. Res.*, 94:1303.

Holzworth, R.H., 1987, Electric fields in the middle atmosphere, *Physica Scripta,* T18:298.

Hughes, T.J., and Rostoker, G., 1979, A comprehensive model current system for high latitude magnetic activity I. The steady state system, *Geophys. J. Roy. Astron. Soc.,* 58:525.

Hunten, D.M., 1955, Some photometric observations of auroral spectra, *J. Atmos. Terr. Phys.*, 7:141.

Kamide, Y., and Matsushita, S., 1979, Simulation studies of ionospheric electric fields and currents in relation to field-aligned currents, 1. Quiet periods, *J. Geophys. Res.*, 84:4083.

Kamide, Y., Richmond, A.D., and Matsushita, S., 1981, Estimation of ionospheric electric field, ionospheric currents, and field aligned currents from ground magnetic records, *J. Geophys. Res.,* 86:801.

Kelley, M.C., Pfaff, R., Baker, K.D., Ulwick, J.C., Livingstone, R., Rino, C., and Tsunoda, R., 1982, Simultaneous rocket probe and radar measurements of equatorial spread F-transitional and short wavelength results, *J. Geophys. Res.,* 87:1575.

Kelley, M.C., Siefring, C.L., Pfaff, R.F. Kintner, P.M., Larsen, M., Green, R., Holzworth, R.H., Hale, L.C., Mitchell, J.D., and Le Vine, D., 1985, Electric measurements in the atmosphere and the ionosphere over an active thunderstorm, 1. Campaign overview and initial ionospheric results, *J. Geophys. Res.,* 90:9815.

Kisabeth, J.L., and Rostoker, G., 1974, The expansive phase of magnetospheric substorms, 1. Development of the auroral electrojets and auroral arc configuration during a substorm, *J. Geophys. Res.,* 79:972.

Lyons, L.R., and Walterscheid, R.L., 1985, Generation of auroral omega bands by shear instability of the neutral winds, *J. Geophys. Res.,* 90:12, 90:321, 90:329.

Lyons, L.R., De la Beaujardiers, O., Rostoker, G., Murphree, S., and Friis-Christensen, E., 1989, *J. Geophys. Res.,* submitted for publication.

Markson, R., 1979, Atmospheric electricity and the sun-weather problem, "Solar-Terrestrial Influences on Weather and Climate", ed. by B.M. McCormac and T.A. Seliga, p. 215, D. Reidel Publ. Co., Dordrecht, Holland.

Meredith, L.H., Gottlieb, M.B., and Van Allen, J.A., 1955, Direct detection of soft radiation above 50 kilometers in the auroral zone, *Phys. Rev.,* 97:201.

Mishin, V.M., Bazarzhapov, A.D., and Shpynev, G.B., 1980, Electric fields and currents in the earth's magnetosphere, in "Dynamics of the Magnetosphere" ed. by S.-I. Akasofu, p. 249, D. Reidel Publ. Co., Hingham, MA.

Mozer, F.S. and Serlin, R., 1969, Magnetospheric electric field measurements with balloons, *J. Geophys. Res.,* 74:4729.

Rees, D., Greenaway, A.H., Gordon, R., McWhirter, I., Charleton, P.J., and Steen, Å., 1984, The Doppler imaging system: initial observations of the auroral thermosphere, *Planet. Space Sci.*, 32:273.

Rostoker, G. and Hron, M.P., 1975, The eastward electrojet in the dawn sector, *Planet. Space Sci.*, 23:1377.

Stern, D.P., 1987, Tail modeling in a stretched magnetosphere, I. Methods and transformations, *J. Geophys. Res.*, 92:4437.

Storey, L.R.O., 1953, An investigation of whistling atmospherics, *Proc. Roy. Soc.* (London), 246:113.

Størmer, C., 1955, "The Polar Aurora", Clarendon Press, Oxford.

Tighe, W.G., and Rostoker, G., 1981, Characteristics of westward travelling surges during magnetospheric substorms, *J. Geoph.*, 50:51.

Tsyganenko, N.A., 1987, Global quantitative models of the geomagnetic field in the cislunar magnetosphere for the different disturbance levels, *Planet. Space Sci.*, 35:1347.

Tsuruda, K., Machida, S., Oguti, T., Kokubun, S., Hayashi, K., and Watanabe, T., 1981, Correlations between the very low frequency chorus and pulsating aurora observed by low light-level television at L = 4.4, *Can. J. Phys.*, 59:1042.

Vallance Jones, A., 1974, "Aurora", D. Reidel Publ. Co., Dordrecht, Holland.

Walker, A.D.M., Greenwald, R.A., Stuart, W.F., and Green, C.A., 1979, STARE radar observations of Pc5 geomagnetic pulsations, *J. Geophys. Res.*, 84:3373.

Wescott, E.M., Stenbaek-Nielsen, H.C., Hallinen, T., Davis, T.N., and Peek, H.M., 1976, The Skylab barium plasma injection experiments 2. Evidence for a double layer, *J. Geophys. Res.*, 81:4495.

Whalen, B.A., Yau, A.W., Creutzberg, F., Gattinger, R.L., Harris, F.R., Vallance Jones, A., McNamara, A.G., Pongratz, M.B., Smith, G.M., Forsyth, P.A., and Koehler, J.A., 1981, Preliminary results from Project Waterhole-an auroral modification experiment, *Can. J. Phys*,. 59:1175.

Winckler, J.R., and Peterson, L., 1957, Large auroral effect on cosmic ray detectors observed at 8 g/cm^2 atmospheric depth, *Phys. Rev.*, 108:903.

LARGE-SCALE ORGANIZATION OF SOLAR SYSTEM PLASMAS

L. J. Lanzerotti

AT &T Bell Laboratories
Murray Hill, New Jersey 07974

ABSTRACT

The solar system is composed of a cellular structure of large-scale plasma envelopes (magnetospheres) embedded within the flowing solar wind, the heliosphere. The latter is itself likely to be a large plasma cell embedded within the local interstellar medium. The heliosphere and magnetosphere plasma cells delineate spatial regions where matter is organized by electromagnetic forces, as opposed to gravitational forces. The recognition of the universality of this concept of the cellular structure of plasmas, both on the large scale of magnetospheres as well as on smaller scales (including self organizing plasma structures), is one of the major developments arising from robotic exploration of the solar system, an exploration which began with the discovery by Van Allen of magnetically-confined radiation around the Earth. This paper discusses the principal attributes and research challenges associated with the existence of the large-scale plasma cells from a personal perspective, with a concentration on the plasma environments of the outer planets and of the Earth. The principal attributes which establish these large-scale cellular structures within the heliosphere include (most importantly) the cell boundaries, the magnetism of the central objects (which are gravitationally-organized matter), the sources of the plasma populations within the cells, and the associated plasma waves - hydromagnetic, electromagnetic, and electrostatic. While a significant synthesis of concepts exists today, substantial work is required for deeper understanding, as well as for intelligent extrapolations of the concepts to astrophysical systems where *in-situ* exploration is not yet possible.

1. INTRODUCTION

The image of the solar system, over the millennia, until the discovery of the Earth's radiation environment by Van Allen [Van Allen et al., 1958; 1983a] in 1958 was that depicted in Figure 1. The Sun and the planets (following their identification or discovery), together with their orbiting moons, as well as cometary interlopers, were considered to be isolated objects in the vacuum of space. Van Allen's monumental discovery, the subsequent investigations of the Earth's space environment by many researchers, and the dispatch of robotic spacecraft to all of the known planets save Pluto, have demonstrated the incompleteness of the depiction in

Magnetospheric Physics, Edited by B. Hultqvist and C.-G. Fälthammar
Plenum Press, New York, 1990

Figure l. Indeed, the organization of matter in the solar system by electromagnetic forces is now recognized to be as pervasive as the organization of matter by gravitational forces. The study of the organization of ionized matter has led to the development of the concept of the "cellular" nature of plasmas in the solar system, as well as to the reasonable speculation that the plasma universe beyond the solar system is permeated by small- and large-scale "cells" of plasmas [Alfvén, 1981]. However, as for the case of the solar system, the electromagnetically-organized plasma environments in the plasma universe are largely unknowable - in effect cells of confined "dark" matter; this contrasts with the gravitationally-organized matter which is frequently observable by its electromagnetic emissions or absorptions.

The recognition of the cellular structures of plasmas in the solar system leads naturally to the pursuit of robotic exploration of the environments. The comparatively easy access of spacecraft to the Earth's plasma environment allows the systematic investigation of it, searching for, and finding, the underlying plasma physical processes which control and organize the dark matter around Earth. Pursuit of knowledge of the plasma environments at the other planets requires substantially greater efforts and resources. (The proximity of the near-Earth plasmas, and even other solar system plasmas, is, however, often a detriment in the competition for resources to carry out space-based research; the more distant the astronomical object, often the more successful is the proposal for research support.)

The organized plasmas found to be associated with the planets and comets of the solar system provide unique opportunities for pursuing scientific understanding. At Earth, organized plasmas exist in the form of large-scale structures such as the plasmasphere and the magnetosphere itself, as well as in the form of small-scale structures, such as night-side auroral current filaments and dayside current vortices (both of which perhaps result in some sense from magnetic field reconnections). Some relatively smaller-scale plasma structures within the Earth's magnetosphere and the heliosphere (such as solar spicules) appear to be of a self-organizing nature. Sub-magnetosphere-size organized plasmas have also been identified

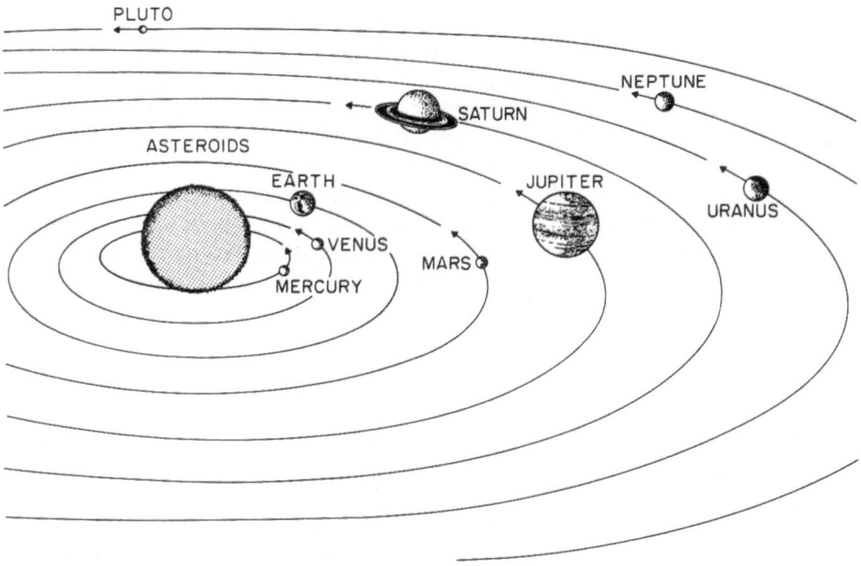

Fig. l. The matter within the planets and of the planets is organized by gravitational forces into the systematics of the solar system. The plasma solar system is nowhere evident in this usual view of the Sun's environment.

within the large-scale plasma cells (magnetospheres) of other planets (e.g., the plasma torus associated with the Jovian moon Io). However, the dominant organized plasma structure recognized at this epoch of exploration of the solar system is that of planetary magnetospheres. Hence, this paper concentrates on the salient features of the large-scale organized plasma cells in the space around the Sun - the planetary magnetospheres and the heliosphere. As shown in other articles in this volume, continued study of the plasmas around Earth will be crucial for future progress in the understanding of electromagnetically-organized matter in the solar system and in the universe. The concentration herein on the large-scale organization of the plasmas is not inconsistent with the intellectual thrust, for example, of galactic astrophysics at the present time, where much attention is being paid to the division of space by gravitational clustering. At the largest scales, these studies examine the three-dimensional distributions of galaxies and their arrangements into bubble-like structures [e.g., de Lapparent et al., 1986; Davis and Peebles, 1983; Bachall and Soneira, 1983; Yoshioka and Ikeuchi, 1989].

In-situ robotic studies of the large-scale plasma cells (magnetospheres and the heliosphere) permit a discernment of the most significant physical attributes that determine their very existence, structures, and stability. Considerations of all of these attributes, from all of the magnetospheres and the heliosphere, ideally would allow as fundamental ranking of large-scale plasma cellular attributes and processes. This, in turn, would allow a more confident extrapolation of theoretical concepts and understanding derived therefrom to the wider universe. Indeed, "the assimilation of large bodies of facts into knowledge and the organization of large bodies of knowledge into comprehensive systems of thought are preconditions", to paraphrase Silber [1989, pg. 67], to the achievement of understanding.

Almost countless research and review articles, and monographs and texts, have been written about the Earth's magnetosphere, touching on the most global to the most arcane aspects. Jupiter's magnetosphere, traversed on four separate occasions by U.S. spacecraft, rates many research and review papers, and even a contributed monograph [Dessler, 1983]. Given this existent and vast literature on solar system plasmas, I must necessarily be selective, and thus I concentrate in this article on common features among the large-scale solar system plasma cells that I personally deem crucial for establishing a basis for physical understanding. By omission, therefore, I deem other attributes to be of lesser, or of secondary, importance for establishing first synthesis and fundamental understanding. Given the relative newness of many of the observational results - to say nothing of the concepts - it is quite possible, even likely, that others, from their own perspectives, would arrive at a somewhat different set of considerations and ranking of attributes. This is normal for a vital, forefront research area whose research advances are being made even while this paper is written.

2. KEY ATTRIBUTES

The boundary regions which establish the large-scale cellular plasma structures are the most significant attributes that *in-situ* spacecraft measurements have revealed. It is within these large-scale plasma cells that vastly disparate physical processes can be dominant, resulting in totally different plasma regimes. Without the ability of even very tenuous plasmas to maintain thin (order of a few ion gyroradii) electrical current sheets over vast distances, there would be an eventual total mixing of plasmas from these different sources, with a resulting homogeneous plasma environment. Thus an assessment of the present understanding of the current sheets in the several plasma regimes measured to date is of primary importance.

However, before this assessment is made, we must enquire about the underlying structure of space that forms the basis for the large-scale spatial variations of the plasmas. The solar

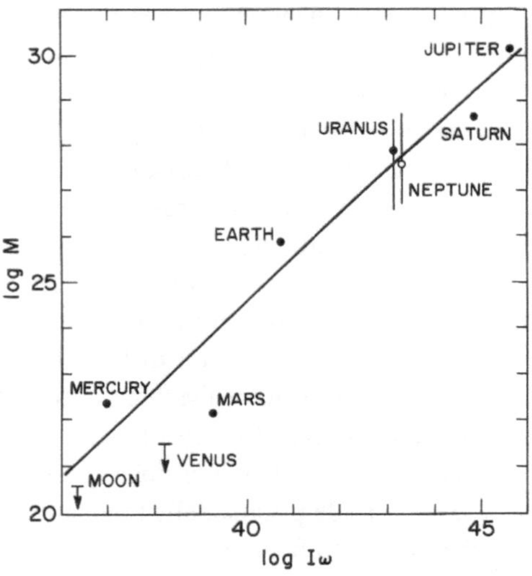

Fig. 2. The magnetic dipole moments of objects in the solar system (in units of Gauss-cm3) organized by their angular momenta. The vertical lines at the locations of Uranus and Saturn were estimates by Van Allen of these value ranges for the two planets prior to Voyager encounter. Adapted from Van Allen (1983b).

system studies have shown the importance, at least in the local environment, of the coupling between the gravitationally-confined matter and the electromagnetically-confined matter. Most of the major gravitationally-confined objects in the solar system have a combination of spin and mass distributions which produce an intrinsic magnetic field. It is this intrinsic magnetic field which forms the underlying basis for the cellular structure of the solar system. However, as will be seen below, the organization of the plasmas often involves more physical considerations than just planetary magnetic fields. [Those solar system objects without intrinsic magnetic fields, such as comets, Venus, possibly Mars, and Titan, often have atmospheres whose ionized outer regions, under the influence of external plasmas, serve to sustain induced currents and magnetic fields.]

Explanations for the origins of the intrinsic magnetic fields associated with rotating, self-gravitating objects in the solar system and the universe represent a major scientific challenge. It is generally accepted that a regenerative dynamo process in the interior of these objects produces an amplification of an initial "seed" magnetic field, probably present during the condensation phase of the matter. Thus, the resultant external magnetic properties are driven by the internal process. The energy for the internal process has been suggested to come from convective thermal motions driven by radioactive isotope decay and heat generation, or a phase change in association with gravitational settling, or a differential precession of interior portions of the object. There has also been a suggestion, originally proposed by Schuster and Blackett, that the magnetic dipole moments of these objects are proportional to their angular momenta.

A recent summary of this hypothesis, due to Van Allen [1983b], is shown in Figure 2. Here the objects' dipole moments, in Gauss-cm^3, are plotted as a function of the objects' angular momenta, Lω. At the time this figure was published, Voyager 2 had not yet reached either Uranus or Neptune. The vertical lines at the locations of these two planets "span ranges of M within which it would not be astonishing to find their actual values" Van Allen [1983b]. The Voyager 2 determination for Uranus is shown by a solid dot while the very recent (August 1989) results for Neptune are shown by the open circle. The Sun also has an intrinsic magnetic dipole field which varies in magnitude in a quasi-periodic, approximately 11 year, fashion.

The dipole moment of the Earth is presently decreasing in value. It is known from paleomagnetic records to have had a range of values over time and to have changed signs many times in the course of Earth's history. Hence, any relationship discerned in Figure 2 must have a time-dependent factor to it. Further, the *in-situ* data from Uranus and Neptune suggests that there is little relationship between the orientation of a planet's spin axis and the magnetic dipole moment, as illustrated in Figure 3. (The situation for Neptune is not shown, but was found to be similar to that for Uranus, providing evidence that the latter is probably not undergoing a magnetic dipole reversal, as was speculated by a number of individuals soon after the measurements were reported). The higher order spherical harmonic terms were found to be more significant for Uranus and Neptune than for some of the other planets (Figure 4), suggesting a possible relationship between the state of condensation, toward the planet's center, of possibly fragmented current-carrying regions and the generation of the body's magnetic field. As an outsider to the field, I have the distinct impression that the *in-situ* robotic spacecraft measurements of the magnetic fields of planetary objects have shown that the state of theoretical understanding of planetary magnetism is not very healthy at present. It certainly represents an area of research with major challenges as well as major opportunities.

3. PLASMA CELL BOUNDARIES

The gravitationally-bound planetary objects, with their magnetic fields, are embedded in plasmas and magnetic fields of solar origin. These plasmas and fields are in motion with respect to the planets. The nominal solar wind velocity V of ~400 km/sec is somewhat more than a factor of ten times the Earth's orbital velocity. The large-scale spatial confinements of plasmas in the solar system arise from the interactions between the flowing solar wind and the planetary magnetic structures. Hence, the underlying attribute of magnetism associated with gravitationally-confined matter is responsible for the occurrence of the most important attribute discovered by robotic spacecraft: the boundaries separating disparate plasma regimes.

(With the full recognition that I might well stimulate disagreement, I do not include here the collisionless shock wave - the bow shock wave - which is formed in front of each solar system object by the super-Alfvénic nature of the solar wind velocity as a key attribute necessary for the synthesis of solar system and astrophysical objects. The discovery and theoretical interpretations of this shock provided new fundamental knowledge to the studies of plasmas. However, the plasma cells around the planets would exist even for a sub-Alfvénic solar wind flow, and such conditions might exist, and even be the dominant condition, throughout the universe.)

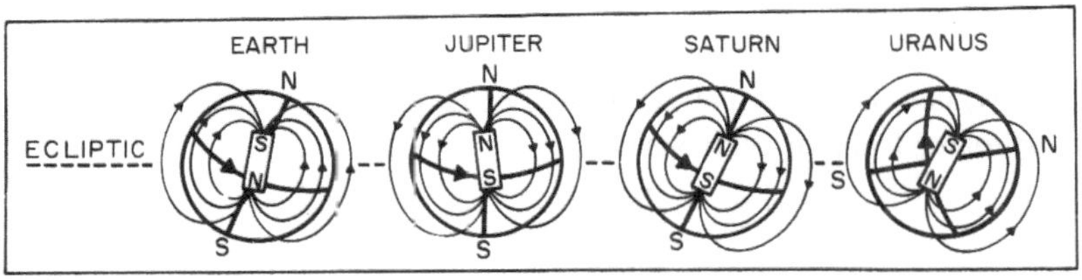

Fig. 3. Orientations of planetary magnetic dipole moments relative to the planets' spin period for Earth and for three outer planets.

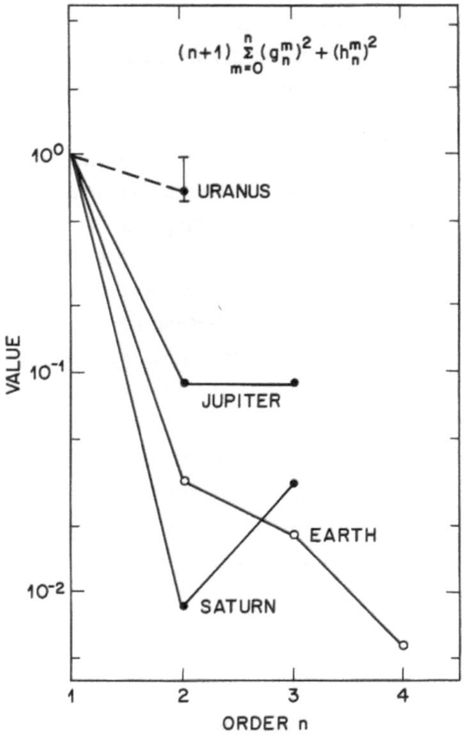

Fig. 4. Relative values of spherical harmonic coefficient fits to magnetic field measurements at the Earth and at the three nearest outer planets. Adapted from Connerney et al. (1987).

As Siscoe [1987] has elegantly explained in a recent review of present knowledge about the Earth's plasma boundary, a boundary exists because the electrically-conducting solar wind plasma is repelled by the planetary magnetic field it finds in its path. The repelling is accomplished by inducing currents in the conductor. These combine with the planet's magnetic field to produce a force which pushes away the conductor. This basic physics understanding led eventually to the macrophysical model of an impenetrable (by particles and fields) magnetospheric boundary (the "magnetopause"). The earliest work in this regard was done by Chapman and Ferraro [e.g., Ferraro, 1952] and the currents which they deduced at the interface between their postulated solar gas streams and the magnetic field are often termed Chapman-Ferraro currents, a term now often extended to currents at boundaries separating solar system plasma regimes.

For the case of a vacuum in the Earth's magnetic field (i.e., the ratio between the external plasma pressure P_{EXT} and the planet's magnetic field pressure is very, very small $\beta = P_{EXT}/(B^2/8\pi) << 1$) extensive work by a number of investigators in the 1960's (see Siscoe [1987] for a historical review) demonstrated that the front boundary location (the sub-solar point) can be rather well estimated from the relations

$$\frac{B_T^2}{8\pi} = P_{EXT} \tag{1}$$

and

$$B_N = 0, \tag{2}$$

where the subscripts T and N denote the confined magnetic field's tangential and normal components, respectively, and

$$P_{EXT} = 2\rho V^2 \cos^2 \psi. \quad (3)$$

Here ψ is the angle between the gas flow and the normal to the boundary ($\cos\psi = 1$ for the subsolar point) and $\rho = nm$ is the mass density. (Siscoe denotes P_{EXT} in (3) as P_D for "Dungey's pressure law," in recognition of the first derivation of it by J. W. Dungey [1958].)

The shape of the Earth's magnetosphere can be calculated in a gas dynamic model, with impenetrable boundaries, for vacuum conditions. However, introducing finite plasma conditions, even for cases of low ($\neq 0$) β, greatly complicates the modelling and is largely beyond the status of measurements for other planets at this time.

In a very general sense, one should balance both plasma and magnetic field pressures on both sides of a cellular plasma boundary. For the case of the solar system, this would be expressed as

$$P_{EXT} + \frac{B_{EXT(T)}}{8\pi} = P_M + \frac{B_T^2}{8\pi} \quad (4)$$

where $B_{EXT(T)}$ is the tangential component of the interplanetary (external) magnetic field, and P_M is the planetary magnetosphere plasma pressure at the boundary. Spacecraft investigations have shown that in the solar system the second term on the left in (4) can always be neglected. However, the P_M term is non-negligible, and even of prime importance for determining the subsolar location of the boundary at Jupiter and Saturn.

The concepts outlined above provide results in crude agreement with *in-situ* spacecraft measurements and lead to the macroscopic scaling of the large-scale solar system plasma cells as shown in Figs. 5a-d. The discoveries at Neptune would, as suggested in the earlier discussion, place it in the same category as the macrostructure of Uranus. Using analogous reasoning, the solar system itself is likely to be a large, three-dimensional plasma cellular structure embedded within the present local interstellar medium (Figure 5d). The validity of this extrapolation to a major astrophysical plasma cell will probably be tested from measurements made by one, or even all, of the four spacecraft presently traveling outward beyond 40 astronomical units (AU), away from the Sun, as well as by the Ulysses spacecraft, which will investigate the polar regions of the Sun after its launch in 1990.

But what of the "real" boundary to the plasma cells? As soon as one inserts plasmas inside the cells (non-vacuum conditions), "non-determinism arises" in the modelling [Siscoe, 1987], not only for the shape of the boundary (of principal concern for the quotation from Siscoe), but also for the characteristics of the boundary region itself. Microphysical plasma processes arise which produce finite current layers and possible breakdowns in the boundary impenetrability. A sketch of some possible (likely) processes at the Earth's magnetopause is shown in Figure 6. Some observational evidence exists for the occurrence of each of the processes, although there is considerable uncertainty as to their simultaneous occurrences and/or their relative importance under various internal and external (relative to the magnetosphere) plasma conditions.

The cell boundary regions which magnetically connect to the "polar" regions of the internal magnetic body are particularly interesting regions with respect to connections between the planet, its plasma environment, and the external plasma environment. Plasma turbulence may facilitate breakdowns in the impenetrability of the boundary in this region. Also, external plasmas can probably penetrate much closer to the central body, even to the surrounding gaseous and/or solid surfaces, than is possible in any other regions.

Fig. 5

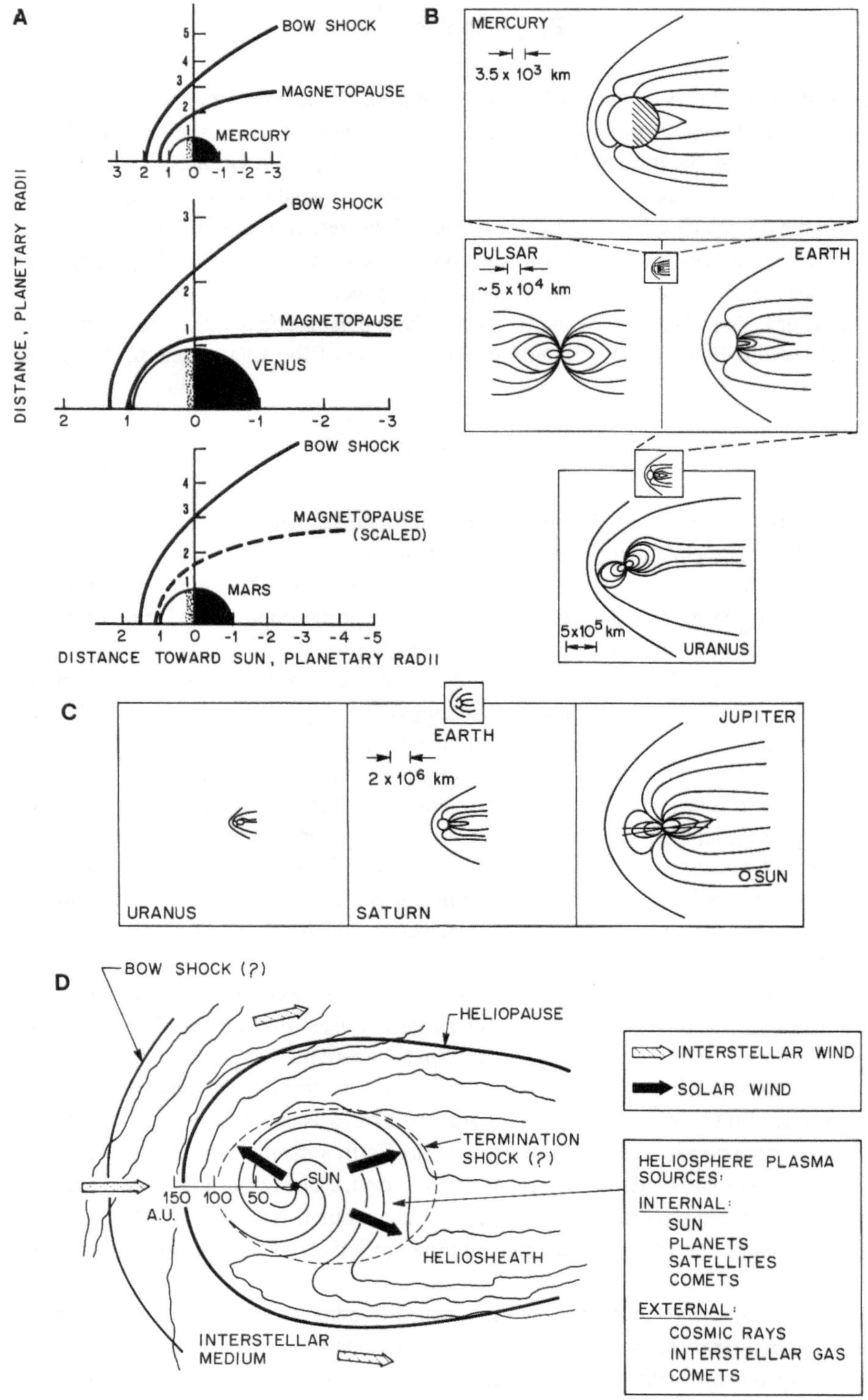
A
BOW SHOCK
MAGNETOPAUSE
MERCURY
BOW SHOCK
MAGNETOPAUSE
VENUS
BOW SHOCK
MAGNETOPAUSE (SCALED)
MARS
DISTANCE, PLANETARY RADII
DISTANCE TOWARD SUN, PLANETARY RADII
B
MERCURY
3.5 x 10^3 km
PULSAR
~5 x 10^4 km
EARTH
5x10^5 km
URANUS
C
EARTH
2 x 10^6 km
JUPITER
SUN
URANUS
SATURN
D
BOW SHOCK (?)
HELIOPAUSE
INTERSTELLAR WIND
SOLAR WIND
TERMINATION SHOCK (?)
SUN
150 100 50
A.U.
HELIOSHEATH
INTERSTELLAR MEDIUM
HELIOSPHERE PLASMA SOURCES:
INTERNAL:
SUN
PLANETS
SATELLITES
COMETS
EXTERNAL:
COSMIC RAYS
INTERSTELLAR GAS
COMETS

The availability of significant computing capabilities has facilitated (encouraged) global modelling of the Earth's magnetosphere and its dynamics [e.g., Walker and Ogino, 1989; Hasegawa and Sato, 1988]. The modelling has the capability of incorporating microscopic plasma processes, at the boundary and internal to the boundary, although the parameter values for these processes (most often the level of dissipation in the plasmas) usually involve *ad hoc* guesses. Hence, intensive work continues at present toward the understanding of plasma and energy transport and dissipation at and across the Earth's magnetopause in both directions.

4. THE PLASMAS

The existence of plasmas in the planetary magnetospheres and in the heliosphere indicates that some aspects of these plasmas must constitute the third significant attribute to be assimilated into a comprehensive synthesis of solar system magnetospheres. But which aspects of the plasmas are of most significance? The temperatures or pressures or compositions or distribution functions or selected energy ranges or? From the perspective of (4), the P_M can be highly significant. Voyager spacecraft measurements in the magnetospheres of Jupiter and Saturn show that near the boundary regions where the spacecraft penetrated and measurements were made, the plasma $\beta \sim 1$ [Krimigis et al., 1979; 1981; Maclennan et al., 1983]. The high values of the plasma pressures were produced by energetic ions with rather high temperatures, as shown from the two example distributions shown in Figure 7. The equality of the internal (magnetospheric) magnetic and plasma pressures at the boundaries suggests that significant loss of internal plasmas to the interplanetary medium can occur [e.g., Krimigis et al., 1979; Zwickl et al., 1981]. The high internal plasma pressures at the magnetopause boundaries of these two large planets means that the boundary structures must be highly dynamic in space and time. The high internal plasma pressures, at least in the case of Jupiter, cause the planet's plasmas to be largely confined to the magnetic equatorial plane by the centrifugal forces exerted by the plasmas on the confining magnetic fields.

Since the internal plasma conditions are important for the structure and stability of the cellular structures, a key plasma attribute to ascertain is the source of these internal plasmas. And is there any conclusion about the source(s) common among all plasma environments? The source(s) of Earth's plasma has been a continuing area of controversy since Van Allen's discovery of the radiation belts. The principal observations have been measurements of the spatial (and temporal) distributions of the plasmas, their energy distributions, and their composition. Observational evidence and theoretical considerations have oscillated over the years between predominantly external and predominantly internal origins. The weight of evidence at present favors the internal - ionospheric/atmospheric - source [e.g., Chappell et al., 1987].

The data at present favor internal sources for the plasma contents of the magnetospheres of the four giant outer planets as well. The moon Io contributes significant low energy plasma

Fig. 5a. Sketch of the nominal location in planetary radii of the bow shock waves and plasma cell boundaries for three of the four "terrestrial" planets.

Fig. 5b. Comparison of the relative sizes of the plasma cells around three solar system planets and the theoretical expectations for a pulsar. The dimension of the magnetosphere of Neptune has been found to be similar to that of Uranus.

Fig. 5c. Comparison of the dimensions of the magnetospheres of three outer planets to that of the Earth.

Fig. 5d. A two-dimensional schematic of the possible plasma cellular structure of the solar system. The interstellar medium can be considered to be moving relative to the solar system because of the latter's proper motion in the galactic disk. Some sources of the plasmas found in the heliosphere are indicated.

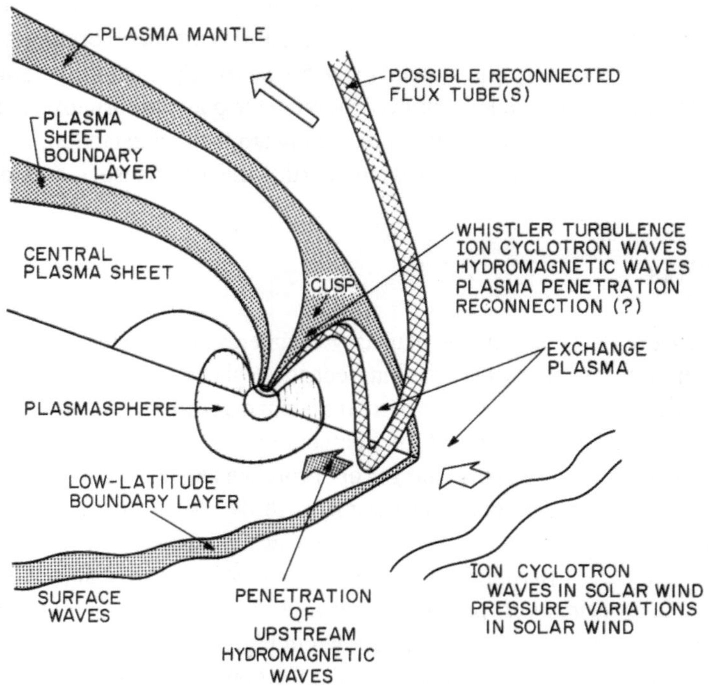

Fig. 6. Schematic illustration of some plasma physical processes operative at the boundary of the Earth's magnetosphere.

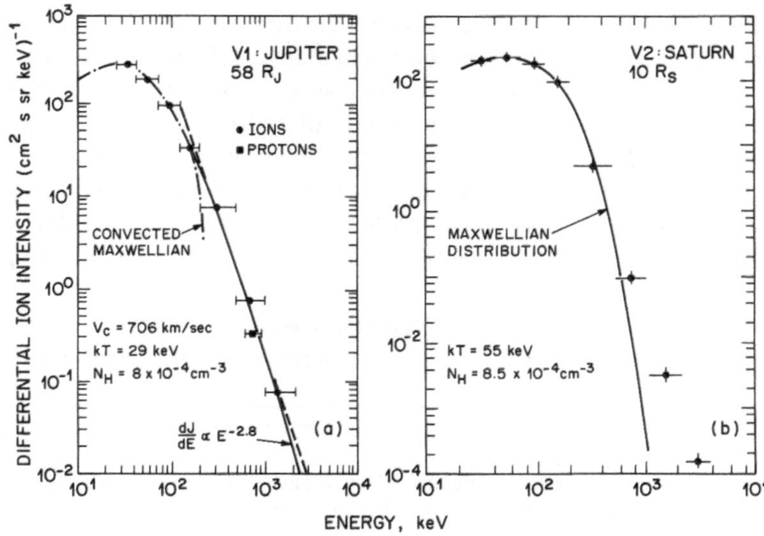

Fig. 7. Hot plasma distribution functions measured by instrumentation on the Voyager 1 and 2 spacecraft within the plasma envelope at Jupiter and Saturn. Hot, tenuous plasmas were found in both planets' magnetospheres (adapted from Krimigis et al., 1979; 1981).

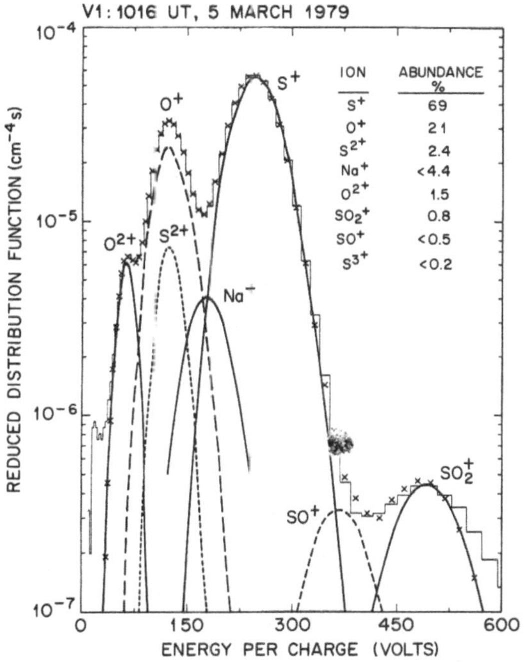

Fig. 8. Distribution function for warm plasmas measured near the orbit of Jupiter's moon Io during the Voyager 1 fly-by of the planet (adapted from Belcher, 1983). The importance of the chemical composition of Io to the magnetosphere plasmas population is clearly evident.

to the inner Jovian magnetosphere, as the distribution functions in Fig. 8 from the Voyager plasma experiment show [Belcher, 1983]. Energetic oxygen and sulfur ions were also observed by Voyager at Jupiter [Hamilton et al., 1981], with spatial distributions consistent with an internal source and abundance levels inconsistent with a solar origin. Energetic oxygen ions at Saturn were attributed to icy moons - a predominantly internal source. The abundances of energetic H_2 and H_3 measured at Jupiter and Saturn (Figure 9), and the abundances of energetic H_2 found in the magnetospheres of Uranus [Mauk et al., 1987] and Neptune [Krimigis et al., 1989], argue for planetary atmospheric origins of these species.

Thus, I would conclude at this time that the major plasma cells around the planets Earth, Jupiter, Saturn, Uranus and Neptune are populated primarily by internal sources, although these sources can be quite different from planet to planet. For Earth, Uranus, and Neptune, the major internal source is the planetary atmosphere, while for Jupiter and Saturn the sources are the planetary atmospheres and the moons, with the moons' contributions being from the interiors ("volcanism" in a broad sense), surfaces, and atmospheres.

The Sun's plasma cell, the heliosphere, also is populated largely internally, as the boxed summary in Figure 5d indicates. The Sun itself, by emitting the solar wind and solar energetic particles, supplies plasmas to the entire heliosphere. The planets, and their satellites, are not insignificant sources, locally, at times, as noted earlier in the discussion of the stability of Jupiter's boundary. The relativistic electron population of the heliosphere is dominated by Jupiter as a source, as the data of Figure 10, acquired by Pioneer 10, shows dramatically. The number of electrons N decreases strongly with increasing radial distance R_J from Jupiter

$$\frac{dN}{dR_J} < 0. \tag{5}$$

The principal external plasma particle sources for the heliosphere are the very high energy (order GeV) cosmic rays, which penetrate directly into the solar domain, and interstellar neutral gases which become ionized and then energized within the solar plasma envelope.

The spatial distributions of the more energetic populations of planetary plasmas significantly depend upon the size of the plasma cell organized around the planet, as given by (4), as well as on the distributions of the cold plasma populations. An interesting initial attempt at a synthesis for a portion of these populations was provided by Cheng et al. [1987] and is shown in Figure 11. Here there is evidence that the phase space densities of energetic (~ 100 MeV/G) ions at the four planets illustrated have a relationship to the cold plasma populations in these planets' magnetospheres. That is, since the particle distribution function F has the property

$$\frac{dF}{dR_P} > 0 \tag{6}$$

for the range of R_p (planetary radii) shown, the source(s) of the energetic ions are at R_p greater than those shown. The spatial gradients in the cold plasma distributions are locales of significant particle loss and plasma wave generation (see below). Hence, details of the plasma distributions must be determined (measured) by *in situ* spacecraft in order to ascertain the conditions under which plasma populations can be generated which will yield observables (e.g., non-thermal "radio" waves) at a distance removed from the planet (see next section).

In the range of planetary radii shown, the cold plasma at Earth originate from the ionosphere (and are contained within a smaller plasma cell, the plasmasphere). At Jupiter, the cold plasmas originate from the moon Io (the Io plasma torus); at Saturn, from the icy rings and moons; and at Uranus, probably from the ionosphere. It should be emphasized that vastly different spatial scales are implied by the horizontal "planetary radii" scale of Figure 11: the radii of Jupiter and Saturn are both ≥10 times the radius of the Earth, while 1 $R_U \sim 5\ R_E$. Further, the fraction of the overall magnetosphere shown in Fig. 11 is dramatically different for each of the planets. Data for about one half of the magnetospheres of Earth, Saturn, and Uranus are shown, whereas the data represent only about 10% to 20% of Jupiter's plasma envelope.

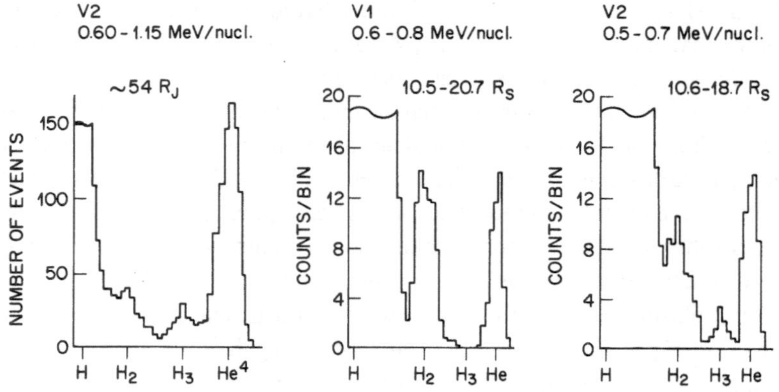

Fig. 9. Energetic hydrogen and helium abundances determined by in-situ measurements on the Voyager 1 and 2 spacecraft at Jupiter and Saturn. The presence of H_2 and H_3 signify planetary atmospheric sources for these species. Figure adapted from Krimigis et al., (1979) and Hamilton et al., (1981; 1983).

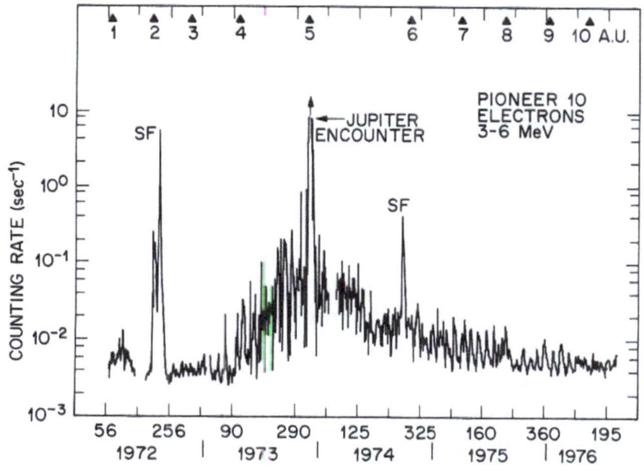

Fig. 10. Relativistic electrons measured in the heliosphere prior to, and after, encounter of the Pioneer 10 spacecraft with Jupiter. The radial distance of Pioneer 10 from the Sun is indicated at the top of the figure. Instances of solar flare (SF) events are noted. The electron gradient, negative away from the planet, is indicative of Jupiter as the source of these high energy heliospheric particles (adapted from Pyle and Simpson, 1977).

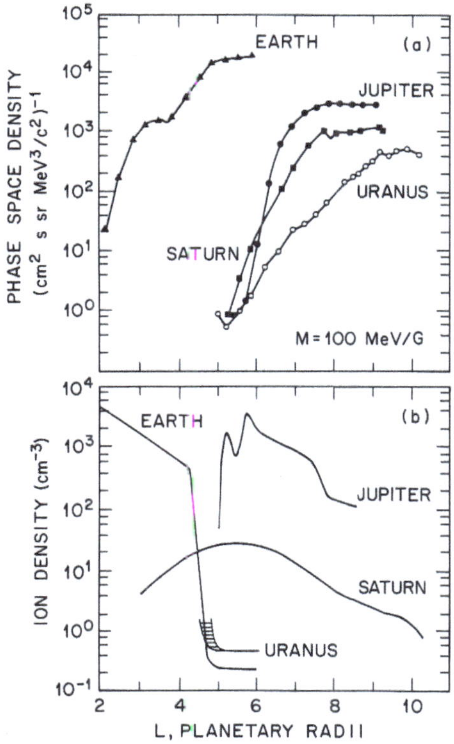

Fig. 11. Top panel: Ion phase space densities as a function of planetary radii for Earth and for three of the outer giant planets. Lower panel: Cold plasma ion densities as a function of planetary radii for the same planets as in the top panel. Adapted from Cheng et al., (1987).

5. THE WAVE ENVIRONMENTS

Another fundamental attribute of the planets' plasma environments are the wave populations. The exchange of energy between the waves and the plasmas establishes the underlying physics that must be included in microscopic and macroscopic models of the plasmas. The existence of information on many aspects of plasma waves in the outer planet magnetospheres is due largely to the persistence and tenacity of the late Fred Scarf, who persuasively advocated the importance of wave measurements on a mission of exploration and discovery such as Voyager. Figure 12 provides a summary overview of the plasma wave environment at Jupiter. In addition to electromagnetic and electrostatic plasma waves and broad-band "noise", plasma hydromagnetic waves (Alfvén waves) are ubiquitous in the magnetospheres of Earth and Jupiter, and probably in other planetary magnetospheres as well. Alfvén waves are also found to propagate in the heliosphere, largely outward from the sun.

The plasma environment of Jupiter is sufficiently energetic that powerful non-thermal waves at decametric and kilometric frequencies are generated and escape the planet's plasma envelope. The intensities of the waves and the relative proximity of Jupiter to Earth (as compared to the other three, more distant, planets) allowed the detection of these waves at Earth by Burke and Franklin in 1954, prior to the space era. The initial, and brief, speculation that these emissions were the result of planetary lightning soon gave way, particularly following Van Allen's discovery of the Earth's trapped radiation, to the understanding that the intense, non-thermal emissions are synchrotron radiation from a trapped plasma population.

The plasma processes which provide the basis for plasma particle energizations and losses, including wave emissions, are hidden from detection by anything other than *in situ* probing of a particular magnetosphere. This is illustrated schematically for Jupiter in Figure 13; where it is seen that the continuum electromagnetic radiation which controls, and is controlled by, the plasma, is trapped in a "potential well" defined by the proton plasma frequency f_p. Alfvén waves are also confined within the planetary magnetospheres. Hence, without direct measurements of the wave fields and the plasmas in a given plasma cell, one cannot determine the underlying physical processes which produce the plasma populations that can radiate at

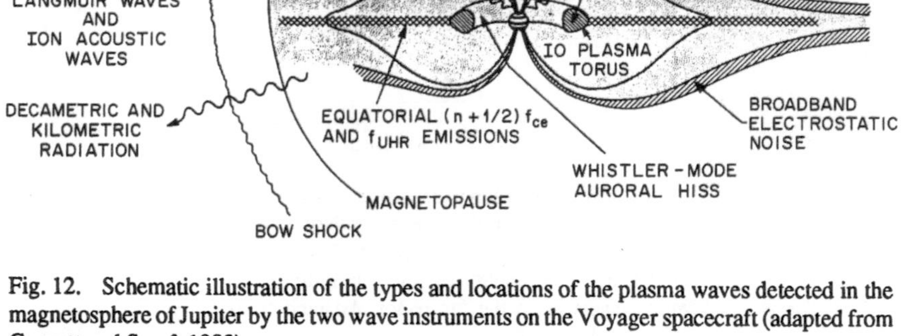

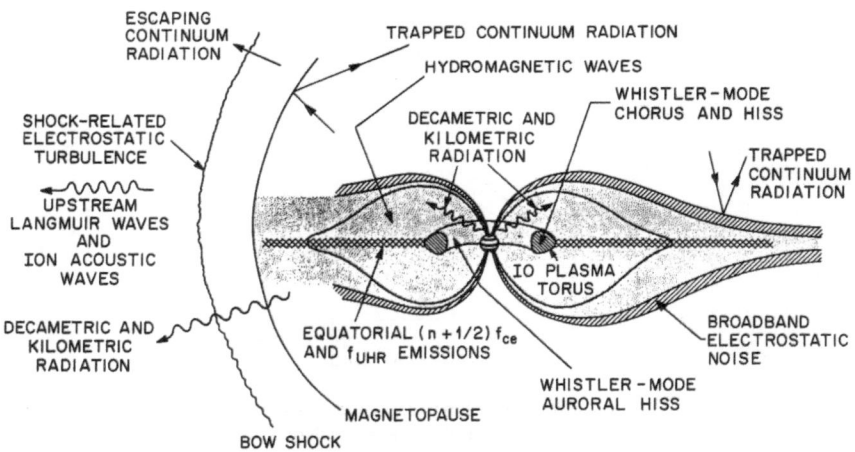

Fig. 12. Schematic illustration of the types and locations of the plasma waves detected in the magnetosphere of Jupiter by the two wave instruments on the Voyager spacecraft (adapted from Gurnett and Scarf, 1983).

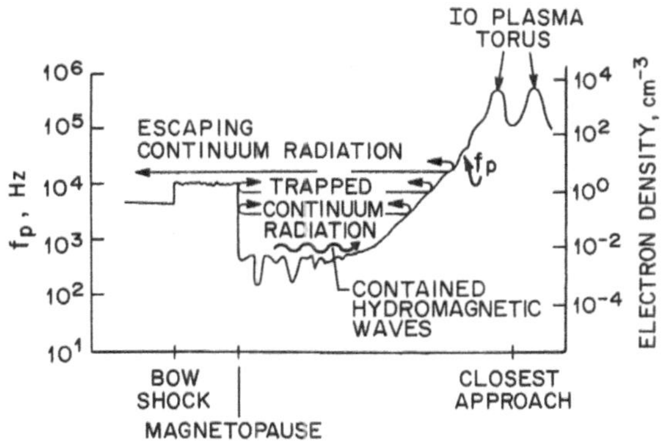

Fig. 13. Schematic illustration of the containment of trapped plasma ("continuum") radiation within the plasma cell around Jupiter. In the left-hand panel, the proton plasma frequency on the left-hand scale is calculated from the electron density determinations (right- hand scale). Adapted from Gurnett and Scarf (1983).

frequencies which can escape. There is little hope, without *in situ* measurements, of achieving satisfactory physical understanding of the large-scale solar system plasma cells.

Only little synthesis exists as yet in comparisons of wave power levels among the planets. Figure 14 illustrates one such comparison of wave energies at several locations in the Earth's and Jupiter's plasma cells. Over-all, the giant planet's wave intensity levels far exceed those produced in the Earth's environment. The wave frequencies at Earth tend to be higher than those at Jupiter when measurements are made within similarly designated plasma regimes.

As emphasized above, the non-thermal plasma waves which escape the plasma envelopes of the planets (and of all remotely-sensed astrophysical objects) are the only evidence which

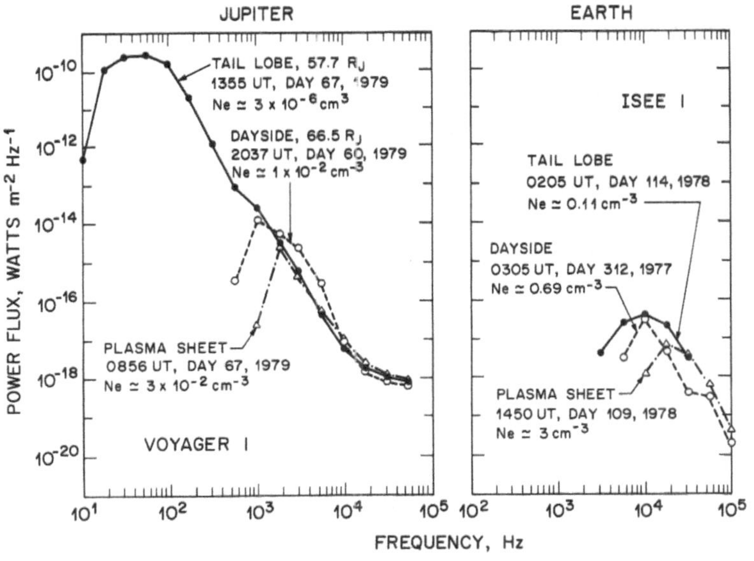

Fig. 14. Intensity levels (power flux) of plasma waves measured in similar plasma regimes in the magnetospheres of Jupiter and of Earth by instrumented robotic spacecraft. Adapted from Gurnett and Scarf (1983).

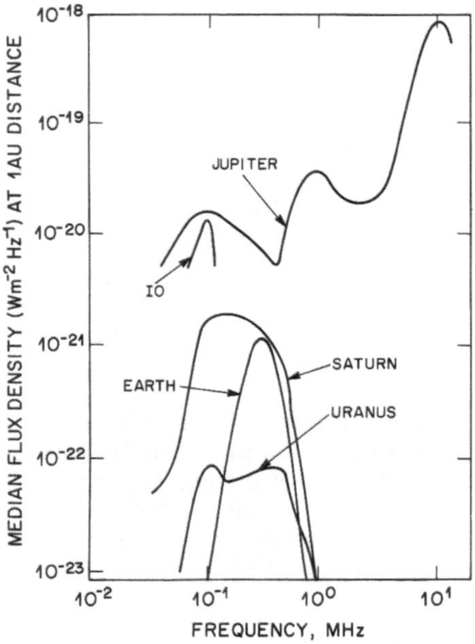

Fig. 15. Median flux densities of radio waves from the Earth and from three outer planets normalized to a distance of 1 AU from each planet. Data courtesy of Dr. M. Kaiser, NASA Goddard Space Flight Center.

we can have, without *in-situ* measurements, of the electromagnetic organization of matter in the universe. A comparison of emitted wave intensities from the planets as a function of frequency and scaled to a uniform detection distance of 1 AU from each planet is shown in Figure 15 (courtesy of M. Kaiser). The results from Neptune, while still under analysis, are similar in intensity to those for Uranus.

Obviously, one of the most important syntheses that must be carried out on the data ("facts") obtained from the *in situ* measurements of the plasmas and waves surrounding the planets is the direct derivation of data such as that contained in Figure 15 from the information on the internal cellular plasmas. Such a derivation is a significant challenge, since even the mechanisms for producing some of the most intense wave emissions from Earth - the auroral kilometric radiation (AKR) - are as yet a subject of intensive investigation and debate.

SUMMARY

The discovery three decades ago by Van Allen of the Earth's plasma environment led to a revolution in concept of the plasma nature of the solar system - and probably of the galaxy and of the universe. *In situ* robotic spacecraft measurements in those accessible regions of the local space environment are crucial for deriving physical understanding of the environment and for confidence-building in making extrapolations of these understandings to non-accessible astrophysical regions.

ACKNOWLEDGEMENTS

I thank G. Haerendel, C. F. Kennel, J. A. Van Allen, S. M. Krimigis, B. Hultqvist, J. G. Roederer, R. Bonnet, J. R. Burch, and C. R. Chappell for very stimulating comments made on this topic at the Crafoord Symposium, C. G. Maclennan, T. Eastman, N. F. Ness, and D. Bryant for helpful comments on the manuscript. The first draft of this paper was composed overlooking the deep blue Tyrrhenian Sea at Torre Normanna, Altavilla Milicia, Sicily. The setting was a constant reminder of the small and large scale beauty of Nature.

REFERENCES

Alfvén, H., 1981, Cosmic Plasma, D. Reidel Pub. Co., Dordrecht, The Netherlands.

Bachall, N. A., and Soneira, R. A., 1983, The spatial correlation function of rich clusters of galaxies, *Ap. J.,* 270:20.

Belcher, J. W., 1983, The low-energy plasma in the Jovian magnetosphere, *in* Physics of the Jovian Magnetosphere, ed. A. J. Dessler, Cambridge Univ. Press, Cambridge, U.K., p. 68.

Chappell, C. R., Moore, T. E., and Waite, J. H., Jr., 1987, The ionosphere as a fully adequate source of plasma for the Earth's magnetosphere, *J. Geophys. Res.*, 92:5896.

Cheng, A. F., Krimigis, S. M., Mauk, B. H., Keath, E. P., Maclennan, C. G., Lanzerotti, L. J., Paonessa, M. T., and Armstrong, T. P., 1987, Energetic ion and electron phase space densities in the magnetosphere of Uranus, *J. Geophys. Res.,* 92:15315.

Connerney, J. E. P., Acuna, M. H., and Ness, N. F., 1987, The magnetic field of Uranus, *J. Geophys. Res.,* 92:15329.

Davis, M., and Peebles, P. J. E., 1983, A survey of galaxy redshift, V. The two-point position and velocity correlations, *Ap. J.,* 267:465.

de Lapparent, Y., Geller, M. J., and Huchra, J.P., 1986, A slice of the universe, *Ap. J.* (Letters), 302: Ll.

Dessler, A. J., (Ed.), 1983, Physics of the Jovian Magnetosphere, Cambridge Univ. Press, Cambridge, U.K.

Dungey, J. W., 1958, Cosmic Electrodynamics, Cambridge Univ. Press, Cambridge, U.K.

Ferraro, V. C. A., 1952, On the theory of the first phase of a geomagnetic storm: a new illustrative calculation based on an idealized (plane not cylindrical) model field distribution, *J. Geophys. Rev.,* 87:2108.

Gurnett, D. A., and Scarf, F. L., Plasma waves, *in* Physics of the Jovian Magnetosphere, ed. A. J. Dessler, Cambridge Univ. Press, Cambridge, U.K., p. 285.

Hamilton, D. C., Gloeckler, G., Krimigis, S. M., and Lanzerotti, L. J., 1981, Composition of nonthermal ions in the Jovian magnetosphere, *J. Geophys. Res.,* 86:8301.

Hamilton, D. C., Brown, D. C., Gloeckler, G., and Axford, W. I., 1983, Energetic atomic and molecular ions in Saturn's magnetosphere, *J. Geophys. Res.,* 88:8905.

Hasegawa, A., and Sato, T., 1989, Space Plasma Physics, l. Stationary Processes, Springer-Verlag, Heidelberg.

Krimigis, S. M., Armstrong, T. P., Axford, W. I., Bostrom, C. O., Fan, C. Y., Gloeckler, G., and Lanzerotti, L. J., 1979, Low-energy charged particle environment at Jupiter: A first look, *Science,* 204:988.

Krimigis, S. M., Carbary, J. F., Keath, E. P., Bostrom, C. O., Axford, W. I., Fan, C. Y., Gloeckler, G., Lanzerotti, L. J. and Armstrong, T. P., 1981, Characteristics of hot plasma in the Jovian magnetosphere: Results from the Voyager spacecraft, *J. Geophys. Res.,* 86:8227.

Krimigis, S. M., and Roelof, E. C., 1983, Low-energy particle population, *in* Physics of the Jovian Magnetosphere, ed. A. J. Dessler, Cambridge University Press, Cambridge, U.K., pg. 106.

Krimigis, S. M., Armstrong, T. P., Axford, W. I., Bostrom, C. O., Cheng, A. F., Gloeckler, G., Hamilton, D. C., Keath, E. P., Lanzerotti, L. J., Mauk, B. H., and Van Allen, J. A., 1989, Hot plasma and energetic particles in Neptune's magnetosphere, *Science,* 246:1483.

Maclennan, C. G., Lanzerotti, L. J., Krimigis, S. M., and Lepping, R. P., 1983, Low-energy particles at the bow shock, magnetopause, and outer magnetosphere of Saturn, *J. Geophys. Res.,* 88:8817.

Mauk, B. H., Krimigis, S. M., Keath, E. P., Cheng, A. F., Armstrong, T. P., Lanzerotti, L. J., Gloeckler, G., and Hamilton, D. C., 1987, The hot plasma and radiation environment of the Uranian magnetosphere, *J. Geophys. Res.,* 92:15283.

Pyle, K. R., and Simpson, J. A., 1977, The Jovian relativistic electron distribution in interplanetary space from 1 to 11 AU: Evidence for a continuously emitting "point" source, *Ap. J.,* 215:L89.

Silber, J. R., 1989, Straight Shooting, Harper and Row, New York.

Siscoe, G. L., 1987, The magnetospheric boundary, *in* Physics of Space Plasmas, Scientific Publ., Inc., Cambridge, Mass, p. 3.

Van Allen, J. A., Ludwig, G. H., Ray, E. C., and McIlwain, C. E., 1958, Observation of high intensity radiation by satellites 1958 alpha and gamma [Explorers I and II], Jet Propulsion:588.

Van Allen, J. A., 1983a, Origins of Magnetospheric Research, Smithsonian Institution Press, Washington, D.C.

Van Allen, J. A., 1983b, Magnetospheres of the outer planets, *in* Essays in Space Science, ed. R. Ramaty, T. L. Cline, and J. F. Ornes, NASA Conf. Publ. 2464, p. 1.

Walker, R. J., and Ogino, T., 1989, Global magnetohydrodynamic simulations of the magnetosphere, *IEEE Trans. Plasma Sci.,* 17:135.

Yoshioka, S., and Ikeuchi, S., 1989, The large-scale structure of the universe and the division of space, *Ap. J.,* 341:16.

Zwickl, R. D., Krimigis, S. M., Carbary, J. F., Keath, E. P., Armstrong, T. P., Hamilton, D. C., and Gloeckler, G., 1981, Energetic particle events (≥30 keV) of Jovian origin observed by Voyager 1 and 2 in interplanetary space, *J. Geophys. Res.,* 86:8125.

THE ROLE OF PLASMA THEORY IN SPACE RESEARCH

C.F. Kennel, F. V. Coroniti, and R. Pellat

Department of Physics
University of California,
Los Angeles, Ca., 90024, USA

ABSTRACT

The status of several "paradigm" problems of space plasma physics will be reviewed as a way to illustrate what role plasma theory has played in space research.

1. INTRODUCTION

In 1978, the US National Academy of Sciences issued a report whose subsequent influence on theoretical space plasma physics in the United States was substantial. Its formal title was *Space Plasma Physics- the Study of Solar System Plasmas,* but people call it the Colgate report, after its chairman, Stirling A. Colgate of the Los Alamos National Laboratory. Colgate's panel of plasma physicists, space physicists, and astrophysicists concluded that:

"Space Plasma Physics is intrinsically an important branch of science. The intellectual significance of the study of solar-system plasmas is documented by its contributions to the development of general plasma physics and by its role in illuminating astrophysical phenomena both internal and external to our solar system."

"On the directly practical side, a better understanding of solar-system plasmas might have substantial importance for terrestrial communications and meteorology."

They expressed their view of the role of plasma theory in space research as follows:

"Now that the initial exploratory stage of space plasma physics has been completed successfully, the fruitfulness of future projects will depend on addressing basic scientific problems. The solution to these problems will call for a logical cycle of theoretical problems definition, the planning of experiments and hence missions, data collection, data reduction, and theoretical analysis, leading to a progressive refinement of the science."

It would be difficult to improve on their formulation.

2. THE COLGATE PROBLEMS

We theorists have a special, but by no means the sole, responsibility to define and clarify "paradigm problems"—the network of facts, theorems, techniques, and beliefs that we use to

Magnetospheric Physics, Edited by B. Hultqvist and C.-G. Fälthammar
Plenum Press, New York, 1990

organize our phenomenology, to place our research in a broad scientific perspective, to motivate sustained research programs, and to judge the progress of our discipline. The Colgate panel found six such problems in space plasma physics:

"We have identified six general abstract problems, vital to further understanding of space plasmas, that have already received considerable theoretical attention and have important implications beyond the study of solar-system plasmas. These are: (1) magnetic-field reconnection, (2) the interaction of turbulence with magnetic fields, (3) the behavior of large-scale flows of plasma and their interaction with each other and with magnetic and gravitational fields, (4) acceleration of energetic particles, (5) particle confinement and transport, and (6) collisionless shocks."

In this paper, we will review how three of the Colgate problems, pitch-angle scattering and spatial transport in planetary radiation belts (5), reconnection in the geomagnetic tail (1), and cosmic ray acceleration (4) have fared in the generation since James A. Van Allen discovered the earth's radiation belts.

3. PITCH-ANGLE SCATTERING AND SPATIAL TRANSPORT OF TRAPPED PARTICLES

We begin by reviewing the oldest microscopic turbulence problem in space plasma physics—pitch-angle scattering by plasma waves. The first experiments on particle radiation trapped in laboratory mirror devices and those we celebrate today, in the earth's radiation belts, took place in the late 1950's, and similar explanations for the anomalous losses of their ostensibly trapped particles emerged soon after (Rosenbluth and Post, 1965, Kennel and Petschek, 1966).

There is one extremely important difference between laboratory mirror devices and the earth's radiation belts. Laboratory devices have a magnetic mirror ratio, M, of about 2, whereas M is $2L^3$ for the geomagnetic dipole field, where L is McIlwain's (1961) L-parameter. In the absence of fluctuations near their cyclotron frequencies, particles conserve their energies and magnetic moments, and their orbits can be classified by their pitch-angle α_o at the mid- or equatorial plane of the magnetic mirror. Trapped particles with pitch angles such that $\sin^2 \alpha_o > 1/M$ bounce back and forth between mirror points, whereas particles with smaller pitch angles, in the so-called "loss-cone", will be lost to the atmosphere or to the end plates of the laboratory device. The fact that the opening angle of the loss cone is about 45 degrees in the laboratory and about 3 degrees at $L = 6$ in the geomagnetic field leads to a crucial difference in behavior when the particles diffuse in pitch angle. Particles which diffuse to the edge of the loss cone in mirror machines are lost on their next bounce. Their "precipitation lifetime" is $1/D$, where D is the pitch angle diffusion coefficient, and the loss cone is virtually empty. On the other hand, a relatively modest diffusion rate can scatter particles into, *and back out of,* the small geomagnetic loss cone in the time, T_B, it takes them to reach the atmosphere. In this case, the pitch angle distribution within the loss cone is nearly isotropic, and the trapping lifetime approaches its theoretical miminum, $T_B M$. We call these two scattering regimes weak and strong pitch angle diffusion, respectively.

By the early 1970's, two classic problems of coupled spatial transport and anomalous loss—similar to those in thermonuclear fusion research—had been defined in magnetospheric physics. The electrons above about 40 KeV energy within the plasmasphere proved to be in weak diffusion, whereas the lower energy, 0.1-10 KeV, electrons precipitating into the diffuse aurora are in strong diffusion, because they are isotropic at low altitudes on auroral field lines. Within the plasmasphere, one balanced the spatial transport due to radial diffusion with losses due to whistler mode pitch angle scattering to calculate the radial, energy, and pitch angle distributions of the electrons in the Van Allen Belts. Beyond the plasmasphere, one balanced the spatial transport to magnetospheric convection with the losses due to pitch angle scattering by electrostatic cyclotron harmonic waves to calculate the radial profiles of the electrons in the diffuse aurora.

3.1 Weak Pitch Angle Diffusion within the Plasmasphere

Although no one knew at the time they were discovered how Van Allen's electrons got their energy, it was striking that they rarely achieved fluxes exceeding about 10^7/cm^2-sec. Kennel and Petschek's (1966) demonstration that this flux corresponded to the threshold for the growth of whistler mode waves which would subsequently pitch-angle scatter the electrons into the atmosphere was one of the first applications of rigorous plasma theory to space observations. It later became clear that their particular theoretical formulation was best suited to the plasmasphere, where a broad-band whistler "hiss"—which satisfies the restrictions on the quasi-linear theory of wave-particle interactions (Kennel and Engelmann 1966)—was a virtually omnipresent feature. It was an obvious theoretical question to ask whether the properties of the radiation belts within the plasmasphere could be understood by imagining that the radiation belt electrons diffused radially across field lines while at the same time the whistler hiss scattered them into the atmosphere.

Lyons (1971, 1972) computed the bounce-averaged electron pitch-angle diffusion coefficients as a function of energy and L-shell, including contributions of all pertinent wave-particle resonances near and far from the equator. He modelled the whistler hiss by a Gaussian frequency distribution centered at 600 Hz, assuming that the wave distribution was relatively isotropic in k-space in the absence of detailed results from ray-tracing analyses. This spectrum was independent of geomagnetic latitude and L-shell. The pitch angle diffusion coefficients were incorporated into the quasi-linear diffusion equation to solve for the electron pitch angle distribution and lifetime as a function of energy and L. These computations reproduced detailed features of the observed pitch angle distributions (Lyons, Kennel, and Thorne, 1972). The precipitation lifetimes due to whistler scattering and Coulomb scattering near the ionospheric mirror points were inserted into the radial diffusion equation to solve for the radial profile of plasmaspheric electrons as a function of energy. Despite the simplifications that were made, the agreement between experiment and theory was quite good (Lyons and Thorne, 1973).

Recent computations continue to confirm the general outlines of the above theory (Solomon *et al.*, 1989), and it seems that our understanding of the transport and loss of radiation belt electrons within the plasmasphere is on a firm conceptual footing. This may not be true for the strong diffusion of diffuse auroral electrons, to which we now turn our attention.

3.2 Strong Diffusion of Diffuse Auroral Electrons and Protons

The spatial morphology of diffuse auroral particles suggests they precipitate from the plasma sheet. The sharp inner edge of the electron plasma sheet appears to map magnetically to the equatorward edge of the diffuse aurora. Precipitating low-energy auroral electrons and protons are often isotropic in pitch angle (Winningham *et al.*, 1975; Evans and Moore, 1979; Sharber, 1981), and their fluxes and energy distributions frequently are virtually identical to those measured on the same field line in deep space (Meng *et al.*, 1979; Schumaker *et al.*, 1989), suggesting that the diffuse auroral electrons are in strong diffusion.

The origin and location of the diffuse aurora was first discussed by Petschek and Kennel (1966) and subsequently by Kennel (1969) and Vasyliunas (1969) independently. Their model involves two fundamental statements. First, earthward convection replenishes the electrons and protons lost from the plasma sheet to the auroral ionosphere. Second, the particles' precipitation lifetimes approach the strong diffusion limit. Combining these two statements has the following consequences. A flux tube deep in the geomagnetic tail hardly loses any particles as it convects towards the earth, because the minimum lifetimes at great distances are very long. Since the convection is thus virtually loss-free, the electron and proton fluxes increase as the flux tube volume decreases. However, the flux tube eventually convects to the point where the flow time and the electron minimum lifetime are comparable, and electrons are rapidly precipitated. The difference between the electron and ion precipitation fluxes must be compensated by a cold return flux of ionospheric electrons (Evans and Moore, 1979); since a hot electron is exchanged for a cold one, the mean energy of the remaining trapped electrons decreases. Thus, this model predicts a sharp decrease of electron temperature at the "inner edge of the electron plasma sheet".

Because protons have longer minimum lifetimes than electrons, they should form their inner edge deeper within the magnetosphere. However, the observed electron and proton precipitation zones more or less coincide, a tell-tale fact that could have served as a warning indicator, but didn't.

Because the cold electron densities beyond the plasmapause are low, whistler mode waves resonate with electrons with energies higher than those observed in the diffuse aurora. Therefore, it was suggested that electrostatic waves with frequencies near odd half-harmonics of the electron cyclotron frequency were responsible for the diffuse auroral precipitation (Kennel *et al.*, 1970). The amplitudes measured by the short antennas on the OGO-5 and IMP 6 spacecraft of the early 1970's were typically 0.1-10 mV/m (Kennel *et al.*, 1970; Scarf *et al.*, 1973; Shaw and Gurnett, 1975). At the same time, Lyons (1974) was finding that a few tenths of a mV/m was sufficient for strong diffusion, so that it appeared that the odd half-harmonic waves could drive the diffuse electron aurora. It was necessary to bounce-average the wave field to relate the electron pitch-angle diffusion coefficient to the measured amplitudes, using an observational estimate of the equatorial localization of the wave amplitude. Estimates closer to 1 mV/m were made later by Belmont *et al.*, (1983) using more modern GEOS results on the equatorial confinement of the waves.

The apparent agreement between the measured and calculated strong diffusion wave amplitudes motivated a long series of investigations into the linear theory of electrostatic cyclotron harmonic waves (Fredricks, 1971; Young *et al.*, 1973; Karpman *et al.*, 1973, 1975; Ashour-Abdalla and Kennel, 1976, 1978; Hubbard and Birmingham, 1978; Kennel and Ashour-Abdalla, 1982). From these analyses came the qualitative realization that cold electrons of ionospheric origin enabled even gentle structures in the electron velocity distribution to be unstable. This linear theory has been tested by simultaneous measurements of waves and resonant particles (Kurth *et al.*, 1979) and by measurements of the wavelengths of odd-harmonic waves near the plasmapause (Filbert and Kellogg, 1988) but unfortunately not in the electron plasma sheet.

Statistical studies of recent measurements using long antennas and more modern receivers indicate that the amplitudes of odd half-harmonic cyclotron waves are rarely high enough for strong diffusion (Gurnett *et al.*, 1979; Belmont *et al.*, 1983; Roeder and Koons, 1989). Yet recent particle data obtained simultaneously near the atmospheric mirror points and near the magnetic equator on substantially the same field line (Schumaker *et al.*, 1989) continue to suggest that plasma sheet electrons are nearly always isotropic near their atmospheric mirror points.

What, if not electron cyclotron harmonic waves, keeps the electron pitch angle distribution isotropic within the loss cone? In which direction should we look to resolve this clearcut paradox? One possibility is that electrostatic cyclotron harmonic waves may not be responsible for the diffuse aurora at all, although, thus far no other candidate plasma wave has been proposed. It might be worthwhile to reexamine whether whistler mode chorus at the inner edge of the electron plasma sheet can scatter at least the higher energy electrons in the diffuse aurora. It is at least conceivable that no plasma waves are needed, if the plasma sheet thins frequently enough to scales where the electron dynamics are locally stochastic (Buchner and Zelenyi, 1987). On the other hand, the electron cyclotron harmonic waves do generally appear to be in the right places at the right times to explain the diffuse aurora. The problem appears to be with their amplitude. Therefore another possibility is that the series of assumptions and computations that enable us to convert a measurement of a wave amplitude at one point on the field line into a bounce averaged diffusion coefficient is faulty. This will require theoreticians to refine the quantitative accuracy of each link in the chain of computations. The full 3-D wavelength spectrum is not measured and the sensitivity as a function of wavelength and antenna orientation of an antenna embedded in a hot magnetized plasma is not completely understood. More refined ray-tracing in a disturbed dipole magnetic field to estimate the fraction of a flux tube that contains intense wave amplitudes might be helpful. (It must be remembered, however, that the wave propagation depends sensitively on the poorly understood properties of the cold electrons on auroral field lines.) It is possible that highly fluctuating amplitudes on the few ms time scale are smoothed in the measurement process, though sample high time-resolution measurements indicate that this is unlikely (Kennel *et al.*, 1970; Belmont *et al.*, 1983). Ground-based optical measurements suggest that the "diffuse" auroral precipitation is highly structured in space and time, so that an equatorial spacecraft measuring wave amplitudes would rarely be in the right place at the right time to detect the waves that tend to isotropize the highly averaged pitch-angle distribution measured by a rapidly moving spacecraft in polar orbit. The theoretical models that relate the pitch-angle scattering rate to the precipitation rate might be flawed; for example the loss-cone size is probably often modified by parallel electric fields. The most fascinating, and probably the most unlikely, possibility is that quasi-linear theory underestimates the pitch-angle scattering rate.

The difficulties with quantitative closure between theory and experiment that we outlined above seem to us to be the most clearcut example of a more general difficulty: the diffusion expected to follow from measured wave amplitudes and the measured diffusion often do not agree. Plasma wave measurements are perhaps the most sensitive diagnostics of changes in plasma state and are very helpful in "encounter physics", in other words, in establishing phenomenology; nonetheless, when wave measurements are used to solve a problem in plasma physics, the quantitative outcome is often in doubt.

If there is no quantitative closure between theory and experiment on the diffuse electron aurora, the very conceptual foundations of the strong diffusion theory of the diffuse ion aurora seem to be faulty. Twenty years of experimental research have turned up no credible candidates for plasma waves that could pitch-angle scatter diffuse auroral ions. Here the problem may lie in the implicit assumption that ions convecting towards the earth conserve their first adiabatic invariant in the geomagnetic tail. Recently, Zelenyi *et al.*, (1989) have suggested that the "stochastic" scattering resulting from the loss of adiabaticity may be responsible for the diffuse ion aurora. If so, no plasma waves would be required.

This review of two problems of transport and loss in the earth's magnetosphere illustrates one primary function of theory: to pose problems simply and clearly enough that the scientific community will work on them in a sustained, evolutionary fashion. Indeed, it took twenty years to complete the first cycle of the perpetual interplay between experiment and theory. Let the second cycle begin.

4. THE CONVECTION-RECONNECTION MODEL OF MAGNETOSPHERIC SUBSTORMS

In the model of magnetospheric convection established by Dungey (1961) and Axford and Hines (1961), the dissipative interaction of the solar wind with the magnetosphere creates a steady, uniform, dawn-to-dusk electric field. The circulation of (low-energy) plasma and of magnetic flux tubes from the nose to the tail of the magnetosphere and back again is described by "**E** x **B**" motions in this steady electric field and the geomagnetic field. This model accounted at once for the circulation of plasma in the polar ionosphere to which the convecting magnetospheric field lines are connected, and explained why the aurora is most intense and active at night, even though the sun, on the dayside, is the ultimate source of the energy for geomagnetic activity.

At the same time the flow pattern was conceived, it was also proposed that magnetospheric convection is driven primarily by the reconnection of the interplanetary and geomagnetic fields at the dayside magnetopause (Dungey, 1961; Levy *et al.*, 1964; Axford *et al.*, 1965; Petschek, 1966; Axford, 1969). The steady reconnection model predicted that open field lines connect the solar wind to the polar ionosphere, so that, as one consequence, the entry of solar wind electrons into the magnetosphere would correlate with the direction of the interplanetary field. It predicted the existence and length of the earth's magnetic tail and the existence of a distant neutral line in the tail which terminates a sheet of hot plasma confined by closed field lines connecting to the nightside auroral oval. The ISEE-3 spacecraft only recently found this neutral line, after twenty odd years in which this most basic requirement of the theory was unsubstantiated. The slow shocks required by Petschek's (1964) MHD theory of reconnection were also discovered at the same time. The fact that people would search for more than a generation for the neutral line and the slow shocks without giving up on the model illustrates how persuasive the original MHD reasoning had been.

The reconnection model demands that the nature and strength of geomagnetic activity depend upon the direction of the interplanetary magnetic field. This further implies that the magnetosphere will be rarely if ever in a steady state, because the solar wind magnetic field is constantly changing direction. Indeed, the complex time-dependences of the activity in the auroral ionosphere were already being arranged into the central phenomenological conception of magnetospheric physics—the auroral substorm (Akasofu,1964)—at about the time the reconnection model was being developed. To the ground-based observer, the auroral substorm starts with a rapid brightening of a small region of the most equatorward auroral arc, which soon spreads to fill much of the visible sky. The challenge has been to relate activity in the ionosphere to motions in the magnetosphere. What the magnetospheric counterpart to the auroral breakup might be was and is still unclear, but the observations and theory led directly to one clear question: could a *time-dependent* reconnection-convection model explain *magnetospheric* substorms?

The first attempt to answer this question was to model what Coroniti (1985b) called later a "conceptual" substorm, one that explains what happens after an initially quiet solar wind field suddenly turns southward and remains southward (Coroniti and Kennel, 1972a,b; 1973; Russell and McPherron, 1973; Schindler, 1974). No real substorm was expected to be like the conceptual substorm, and the model's function was only to outline how the steady convection model might be extended to substorms.

Evidence had been accumulating that a growth phase (McPherron, 1970) of magnetospheric substorms commences with the initiation of strong reconnection at the dayside magnetopause, and terminates with the development of a *new* reconnection region in the near-earth plasma sheet (Hones, 1973, McPherron *et al.*, 1973; Russell and McPherron, 1973). This was the most natural definition of the growth phase in the time-dependent reconnection-convection model. The plasma sheet was expected (and required!) to thin during the growth phase until the conditions for reconnection near the earth were ripe. It was tempting to associate the initiation of tail reconnection with auroral breakup.

It is important to note that subsequent evidence has continued to support the reconnection interpretation of the global tail dynamics of substorms (Hones, 1979; McPherron, 1979; Coroniti *et al.*, 1980). In particular, substorm-generated plasmoids in the distant tail (Hones *et al.*, 1984) are the most easily explained by near-earth reconnection.

The conceptual model used qualitative reasoning and semi-quantitative calculations based on the principles of magnetohydrodynamics to provide a framework for the interpretation of observations. However, it was not absolutely certain that the conceptual substorm was a quantitatively consistent deduction from magnetohydrodynamics. The first global magnetospheric simulations (LeBoeuf *et al.*, 1978, 1981; Lyon *et al.*, 1981; Brecht *et al.*, 1982), however dissipative and poorly resolved spatially, did confirm that the conceptual substorm was based on correct magnetohydrodynamic reasoning. Second-generation simulations added confirming detail. Two- and three-dimensional MHD simulations (Birn, 1980, Birn and Hones, 1981) that focussed on tail reconnection onset yielded plasma sheet flows and tailward plasmoid ejections like those observed. Ogino *et al.*, (1984, 1985, 1986) and Ogino's (1986) studies of field-aligned currents and magnetospheric convection, driven by variations in both the y- and z-components in the interplanetary magnetic field, reproduced the current and flow patterns observed on polar-orbiting spacecraft (Walker and Ogino, 1988).

In short, numerical MHD simulations arguably do account for the large-scale aspects of the dynamics of the magnetosphere and its interaction with the ionosphere (Walker and Ogino, 1989). Thus, the conceptual substorm has evolved seamlessly into the "numerical MHD substorm", which now performs with much greater precision the function of the original steady convection model: to put single-point spacecraft measurements in a consistent phenomenological context.

What, then, is the problem?

The problem is that the plasma sheet is not an MHD system. It is truly collisionless, and its basic scale is an ion Larmor radius, so that MHD cannot be strictly valid, although the hope has been that it will describe the average situation. Moreover, reconnection must violate the ideal MHD approximation, the physics of that violation is still uncertain, and it remains unclear whether kinetic reconnection can do what resistive reconnection does in MHD substorm simulations. One obvious question was whether "anomalous resistance" could develop in the magnetic neutral sheet. Coroniti and Eviatar (1977) found that ion acoustic anomalous resistance is possible probably only in extremely thin neutral sheets. Similarly, the lower hybrid drift instability (Huba *et al.*, 1978, 1980, 1981) cannot make anomalous resistance where it is needed, in the center of the magnetic neutral sheet (Coroniti, 1985a). It now seems unlikely that anomalous resistance can provide that which makes the numerical substorms go.

Difficulties like those above did not seem terribly critical when the reconnection-convection model was first being developed, for a beautiful model of collisionless tearing mode reconnection (Laval *et al.*, 1966) was already at hand to reassure us that reconnection really might occur in the geomagnetic tail (Coppi *et al.*, 1966). Later on, those constructing conceptual substorm models naturally viewed collisionless tearing mode reconnection and its variants as a possible mechanism for substorm breakup (Schindler, 1972, 1972, 1980; Galeev *et al.*, 1978; Coroniti, 1985b).

The original tearing mode calculation (Coppi *et al.*, 1966) was made for a true magnetic neutral sheet with *no* component of the magnetic field normal to the sheet. In a sense, it assumed that reconnection was already occurring. However, the conceptual substorm model requires that reconnection reoccur on *already closed* field lines with a *non-zero* normal component. Schindler (1974) argued that since electrons are easily magnetized by a small normal component of the magnetic field, it would be an ion tearing mode that is responsible for substorm tail reconnection. However, Galeev and Zelenyi (1976) and Lembege and Pellat (1982) showed that electron magnetization strongly stabilizes the ion tearing mode.

There followed numerous attempts to overcome the stabilization of the collisionless tearing mode by magnetized electrons. Coroniti (1980) suggested that electron pitch-angle scattering due to small-scale electromagnetic turbulence would allow the ion tearing mode to grow, but it is not clear that the turbulence observed in the neutral sheet allows it to grow enough to induce a new X-type neutral sheet (Coroniti, 1985). Buchner and Zelenyi (1986, 1989) and Chen and Palmadesso (1986) showed that electron orbits can be nonadiabatic and chaotic in a thin magnetized neutral sheet. Buchner and Zelenyi (1987) suggested that chaotic diffusion of electrons may induce ion tearing near the earth if the plasma sheet thins enough during substorm growth phase. Their calculations suggest that electron stochasticity, and thus tearing, would set in if the current sheet thins to a characteristic scale of 1500 km. It has proven quite difficult to ascertain experimentally whether this condition is achieved at breakup.

Galeev *et al.*, (1978) proposed that a finite amplitude external perturbation could create a new X-type neutral point; once the neutral point is established, ion tearing mode dynamics forces the rate of reconnection to increase explosively. Although explosive ion-tearing model has been simulated numerically (Teresawa, 1981; LeBoeuf *et al.*, 1982), both the theory and the simulations overlooked essential stabilizing features of electron dynamics (Lembege and Pellat, 1982). Coroniti's (1985b) later model of inductive collapse to explosive ion tearing suffers from the same shortcoming.

Electron stabilization of the tearing mode in quasi-neutral sheets simply hasn't gone away, despite many theorist's wishes that it do so. The necessary obsession with making the tearing mode grow has also deferred adequate consideration of other problems. For example, the temporal and spatial relationship between tail reconnection and auroral breakup has not been established experimentally with theoretically decisive accuracy; if they were indeed closely related, the tearing mode would have to explain why breakup starts in a small region typically around local midnight. Nearly all studies of the tearing mode have been two-dimensional, and indications are that localized tearing modes will not grow any more easily than those that extend clear across the geomagnetic tail.

In summary, the tearing mode may not grow in neutral sheets with finite normal magnetic field components. If it can grow, will it grow near the earth, where "dipolarization" of the tail-like magnetic field is the clearest manifestation of substorm breakup? It may not be able to grow fast enough in any case; and it is difficult to see how the tearing mode could produce a breakup that starts locally. All this could add up to a crisis for the tearing mode model of substorm breakup.

The crisis may even extend back to where it all started: to the notion of quasi-steady convection in the earth's magnetic tail at quiet times. The quiet time plasma sheet dynamics reported by Coroniti *et al.*, (1980) did not have a coherent phenomenology. The flow was not steady but often occurred in 10 to 20 minute bursts; these sporadic flows could be earthward, tailward, or downward. When coherent earthward flows were found in the plasma sheet at IMP-7 (35 earth radii), they were not always accompanied by substorms in the ionosphere, suggesting that the flow was localized across the plasma sheet or that it did not approach the earth. Recent high time resolution measurements show that the flow in the central plasma sheet near the earth (where substorm breakup may well start) differs even more radically from the steady convection model. The average flow speed is less than 100 km/s; high-speed flows do occur but in bursts of one minute duration (or less!). These bursts are interspersed with intervals of essentially stagnant plasma flow (Baumjohann *et al.*, 1989).

Perhaps the "quiet state" of the plasma sheet in which substorm tail reconnection is supposed to start should not be pictured as quiet at all. Perhaps irregularity and spatial structure obscure steady convection, or tell us it does not occur except after substorm breakup. All this brings us full circle, back to the diffuse aurora. When it is seen from the ground, the diffuse auroral light is certainly distinguishable from long thin arcs, but this does not mean that the diffuse aurora is smooth (S.Akasofu, private communication, 1989). Indeed, when it is viewed with high time and spatial resolution, the diffuse aurora appears to be organized into independent, intermittent patches of complicated shape (T. Oguti, private communication, 1989). Much of this structure is obscured by our poor eyes, our slow cameras, and the particle detectors carried by our fast-moving polar orbiting spacecraft. It is well-nigh impossible to know with meaningful precision where in the plasma sheet this structure will map. Yet if structure and irregularity prevail both in the diffuse aurora and in the plasma sheet, very detailed statistical characterizations will be needed to connect spacecraft measurements of waves with the precipitation of plasma sheet electrons into the atmosphere.

Perhaps we have been careless in our use of words such as "quiet" and "diffuse"; if so, blame theory.

5. THE PLASMA PHYSICS OF COSMIC RAY ACCELERATION

Cosmic ray acceleration is quite possibly the problem of the most wide-spread significance to which research in solar-system plasmas makes a critical contribution. Its present status reveals much about the level of our understanding of plasma physics.

Fermi was the first to suggest that scattering from moving magnetic irregularities would produce the power law energy distribution characteristic of cosmic rays. The properties of Fermi's moving magnetic clouds had to be carefully tailored to produce the observed spectral index, and nowadays Fermi's clouds have been replaced by large amplitude magnetic turbulence—Alfvén waves. The waves, which act as scattering centers, are set in systematic relative motion relative to one another over a large spatial scale by a collisionless shock. In the shock's frame of reference, the waves upstream are blown towards the shock at highly super-Alfvénic speeds and are then transmitted through the shock in to the downstream sub-Alfvénic flow, where they propagate slowly. Therefore the upstream and downstream wave distributions are in relative motion, and a particle which scatters from the waves back and forth across the shock systematically gains energy. Its net energy gain depends upon the number of times it crosses the shock and not on the wave amplitudes upstream and downstream, so long as the amplitudes are large enough over a large enough distance to reflect a particle back towards the shock. The MHD turbulence can be present in the medium before the shock sweeps over it, but the more attractive possibility is that the turbulence is generated by the energetic particles themselves by the Alfvén Ion Cyclotron instability.

There has evolved a relatively simple picture of Fermi accelerating shocks that in broad outline fits our observations of quasi-parallel shocks in the solar system—those that propagate sufficiently parallel to the upstream magnetic field that shock-heated particles can escape freely into the upstream region from downstream to start the Fermi acceleration process. The overall shock consists of a thin subshock (in which the dissipation is due to the "conventional" microturbulence responsible for "ordinary" collisionless shock structure), and broad foreshock and postshock regions in which energetic particles interact with Alfvén turbulence.

The energetic particles, which are approximately isotropic in the frame of the shock, have a pitch angle anisotropy of the firehose sense in the frame of the upstream flow, and so destabilize Alfvén waves, which amplify as the super-Alfvénic flow carries them towards the subshock. As the waves are transmitted through the subshock, they couple to a variety of MHD modes, and most of this wave energy is convected downstream. If the large amplitude waves upstream and downstream can scatter particles leaving the subshock in either direction back through the subshock, the cycle of upstream and downstream scattering results in a net energy gain. The resulting momentum distribution will be a power law whose spectral index depends only upon the subshock compression ratio, and is comparable to what is needed for galactic cosmic rays. This salient fact, discovered virtually simultaneously by Axford *et al.*, (1977), Bell (1977a, 1977b), Blandford and Ostriker (1978), and Krymsky (1977) at about the time the Colgate panel was completing its report, has encouraged people to work in earnest on the plasma physics of shock acceleration ever since.

The discovery that shock acceleration produces a power law spectrum reinvigorated astrophysical research on the acceleration of cosmic rays by the interaction of supernova shocks with the interstellar medium. Supernova shocks can last some tens of thousands of years before they weaken into insignificance, and the hope is that the particles have enough time to diffuse back and forth across the shock front enough times to achieve the high energies that are observed.

The supernova remnant and its associated shock decelerate as the shock sweeps up interstellar matter, so that for most of its lifetime, the Mach number of the shock is relatively low—comparable with those we encounter in the interplanetary medium. Moreover, most of the volume of the interstellar medium is in the so-called "hot, low-density phase", a fully ionized plasma phase whose temperature is comparable with the solar wind's, but whose density is smaller. It is difficult to measure the magnetic field in the hot low density phase, but it is thought that β, the ratio of thermal to magnetic pressure, is about 10, somewhat larger than is normal for the solar wind, but within the range of variability observed.

The above arguments suggest that it may be profitable, despite the known strong dependence of collisionless shock structure upon plasma parameters, to examine how shocks in the solar system accelerate energetic particles, not only for its own sake, but also to see what the solar system tells us about galactic cosmic ray acceleration. Certainly, solar system shocks are the best hope we have for studying the microphysics of particle acceleration by shocks.

It is not our intention to review the excellent experimental research on particle acceleration by solar system shocks. Moreover, we do not wish to imply that Fermi acceleration is the only way that particles in the solar system are accelerated, or the only way that solar system shocks accelerate particles. However, we must express our belief that the general outlines of the shock acceleration theory, which is highly idealized, are supported, *in the low-energy limit,* by observation. It now seems clear that quasi-parallel interplanetary shocks do have extended foreshocks that contain energetic protons and finite amplitude MHD turbulence. Moreover, planetary bowshocks also have foreshocks upstream of those portions of the curved shock surface where the shock is locally quasi-parallel. A presently active field of research is to test quasi-linear theories of these foreshocks (Lee, 1982, 1983) using spacecraft measurements of the waves and energetic particles.

Kennel (1989) has reviewed three key problems of plasma physics associated with what has now become the "standard" model of cosmic ray acceleration. The first issue is the so-called "seed particle" or efficiency problem. Since the cosmic rays appear to be directly accelerated out of the interstellar medium, we must ask how an originally cold plasma particle begins to participate in the Fermi process. We expect the thermal interstellar plasma to be heated by the collisionless dissipation in the subshock, and that some of these subshock-heated particles will acquire sufficient energy to stream along field lines, catch, and pass through the subshock from behind. When these seed particles get upstream, there must be enough of them to stimulate the growth of the Alfvén waves resonant with them to amplitudes sufficient to reflect them back to the subshock from upstream. They then are subject to Fermi acceleration thereafter.

Only one step in the above chain of processes is reasonably well understood. A subshock-heated or reflected ion with two or three times the downstream thermal speed can easily catch the subshock, provided the subshock propagates at angles less than about 45 degrees to the upstream magnetic field direction. The remaining issues cannot be settled until we have a clearer picture of the microdissipation in quasi-parallel shocks. The rate at which this microstructure creates an escaping superthermal tail in the downstream ion distribution or reflects particles upstream determines how many seed particles are produced and, ultimately, the overall acceleration efficiency.

The second issue is the "reflection problem". Linear theory predicts that an ion with a given component of velocity parallel to the magnetic field will resonate with circularly polarized Alfvén waves according to the well-known cyclotron resonance condition. The quasi-linear theory of ion pitch angle scattering, which retains the linear resonance condition, predicts that the pitch-angle diffusion coefficient will be zero at 90 degrees pitch angle. Thus there is no way in quasi-linear theory to reverse the particle's parallel velocity, as the Fermi mechanism requires.

In short, shock Fermi acceleration is necessarily a strong turbulence process. Simple energy balance arguments indicate that the Alfvén waves will reach order unity amplitude near the subshock, so that we do not expect quasi-linear theory to apply anyhow. Most theorists simply assume that ion reflection occurs and move on, but precisely how this occurs is unclear. It is probably important to grapple with this relatively academic issue, because understanding it may help us to deal with what is an absolutely central problem with acceleration theory which we discuss next.

The third issue, and, in the long run, the most difficult one, concerns how particles diffuse to the extremely high energies observed in the cosmic rays. Using a quasi-linear estimate for the proton scattering rate (but assuming that the reflection problem is somehow solved) Galeev *et al.*, (1986) find that the exponential scalelength of the energetic particles in the foreshock becomes comparable to the radius of curvature of typical supernova remnants at about 3×10^{11} eV, well below the 10^{14} eV to which the cosmic ray spectrum extends without a break in the spectral index. Furthermore there are not enough 10^{14} eV particles to destabilize the very long wavelength waves needed to scatter them. There are a variety of ways to reach this conclusion but the conclusion is always the same: according to the best currently available theory of particle scattering, quasi-linear theory, even an order unity Alfvén wave field cannot scatter particles fast enough to permit them to be accelerated by supernova shocks to the high energies required.

One may choose between two points of view about the above problem. The first is to admit that a few times 10^{11} eV is all you get from supernova shocks and to look for larger shocks of galactic scale to boost the particle energy the remaining two orders of magnitude. On the other hand, the absence of a break in the spectrum suggests that particles are indeed accelerated to 10^{14} eV by the same process in a self-similar way. This second alternative presents by far the more challenging problem. It implies that the scattering rate at high energies is much faster than quasi-linear and is not governed by the linear resonance condition.

Measurements on interplanetary shocks confirm that particles can be directly accelerated out of the thermal background (Gosling *et al.*, 1980), and measurements of the energetic ion diffusion coefficients just upstream of interplanetary shocks are in good agreement with Lee's (1983) quasi-linear theory (Kennel *et al.*, 1986, Tan *et al.*, 1989). However, only the low energy limit of the cosmic ray acceleration theory can be tested by measurements in solar system plasmas. Moreover, present numerical simulations of quasi-parallel shocks in solar system plasmas are able, at best, to resolve the first steps in the Fermi acceleration process. The very high energy limit will be difficult to attack using numerical simulations because of the very large range of spatial scales involved in scattering particles with a spread of 10^5 in Larmor radius.

To study the high energy limit of shock acceleration theory, theoreticians will have to define, and persuade their experimental and simulation colleagues to attack, an agenda of research that is aimed at understanding particle acceleration, not in the solar system, but in astrophysical contexts. We must ask space physicists to answer questions that are not of direct importance to the understanding of solar system plasmas. Can we detect any deviations from the quasi-linear scattering rate in our data on nonlinear Alfvén waves in the solar system? In which direction do these deviations point? The complexity of the solar system environment should also remind astrophysicists of contingent events not contained in their idealized theories of particle acceleration.

Perhaps the single most important objective is to understand the physics of large amplitude Alfvén wave turbulence. In particular, why do the Alfvén waves we observe often have nonlinearly developed waveforms? What is the physics of pitch angle scattering from the nonlinear Alfvén turbulence we observe in the solar wind, near comets, and upstream of shocks? Can the pitch angle scattering rate exceed quasi-linear? These questions have general plasma physical significance, and the solar system is the best place to study them.

The problem of cosmic ray acceleration exemplifies one key function of theory in solar system plasma physics: to relate space plasma problems to analogous ones in astrophysics. Despite the fact that it has not been solved, this problem, which has been addressed by theory, numerical simulations, and spacecraft observations, is probably the most successful example to date in which the interactions between plasma physicists, space physicists, and astrophysicists have brought a major problem into clear focus.

6. CONCLUDING REMARKS

We have reviewed three theoretical problems central to space plasma research: pitch-angle diffusion and the precipitation of trapped particles to the earth's atmosphere, the reconnection-convection model of magnetospheric substorms, and the acceleration of energetic particles by shocks. All three paradigm problems have encountered serious difficulties which some say are life-threatening.

It is truly not clear whether we have made significant progress on the central problems of theoretical space physics, *at the level demanded by the Colgate report,* since that report was completed over a decade ago. One can imagine many reasons why, but in truth, no one really knows. In our view, it is because the fundamentals are not settled. Space research communicates much about the dynamics of collisionless plasma, but the theoretical vocabulary of plasma physics is still too attenuated to decipher our observations and to communicate our understanding in fundamental terms that are comprehensible to the general community of physicists.

ACKNOWLEDGEMENTS

S.-I. Akasofu, J. Buchner, and J. R. Kan participated in framing the arguments in this paper, but are not responsible for any errors of fact, attribution, or interpretation contained in it. CFK would like to thank the Geophysical Institute of the University of Alaska, Fairbanks, for its hospitality, during the time the first draft was composed, and the participants in the

Crafoord symposium, whose spirited remarks contributed much to the final form of this paper. This research was supported by NASA NAGw-1624, and by NSF ATM-8814955.

REFERENCES

Akasofu, S.-I., The development of the auroral substorm, *Planet. Space Sci.*, 12, 273, 1964

Ashour-Abdalla, M. and C. F. Kennel, Convective cold upper hybrid instabilities, in *Magnetospheric Particles and Fields,* edited by B. M. McCormac, p.181, D. Reidel, Dordrecht, Holland, 1976

Ashour-Abdalla, M., and C.F. Kennel, Nonconvective and convective electron cyclotron harmonic instabilities, *J. Geophys. Res.*, 83, 1531, 1978

Axford, W.I., and C.O. Hines, A unifying theory of high-latitude geophysical phenomena and geomagnetic storms, *Can. J. Phys.*, 39, 1433, 1961

Axford. W.I., H.E. Petschek, and G.L. Siscoe, Tail of the magnetosphere, *J. Geophys. Res.*, 70, 1231, 1965

Axford, W. I., Magnetospheric convection, *Revs. Geophys.*, 7, 421, 1969

Axford, W.I., E. Leer, and G. Skadron, The acceleration of cosmic rays by shock waves, in *Proc. 15th Int. Conf. Cosmic Rays,* 11, 132 (1977)

Baumjohann, W., G. Paschmann, and C.A. Cattell, Average plasma properties in the central plasma sheet, *J. Geophys. Res.*, 94, 6597, 1989

Bell, A.R., The acceleration of cosmic rays in shock fronts, 1, *Mon. Not. Roy. Astron. Soc.* 182, 147, (1977a)

Bell, A.R., The acceleration of cosmic rays in shock fronts, 2, *Mon. Not. Roy. Astron. Soc.* 182, 443, (1977b)

Belmont, G., D. Fontaine, and P. Canu, Are equatorial electron cyclotron waves responsible for diffuse auroral precipitation? *J. Geophys. Res.*, 88, 9163, 1983

Birn, J. Computer studies of the dynamic evolution of the geomagnetic tail, *J. Geophys. Res.*, 85, 1214, 1980

Birn, J., and E. W. Hones, Jr., Three-dimensional computer modeling of dynamic reconnection in the geomagnetic tail, *J. Geophys. Res.*, 86, 6802, 1981

Blandford, R., and J.P. Ostriker, Particle acceleration by astrophysical shocks, *Astrophysical Journal,* 221, L29, (1978)

Brecht, S.H., J.G. Lyon, J.A. Fedder, and K. Hain, A time-dependent three-dimensional simulation of the earth's magnetosphere:reconnection events, *J. Geophys. Res.*, 87, 6098, 1982

Buchner, J., and L.M. Zelenyi, ''Deterministic chaos in the dynamics of charged particles near a magnetic field reversal,'' *Phys. Lett. A.*, 118, 395, 1986.

Buchner,J., and L.M. Zelenyi, ''Chaotization of electron motion as the cause of internal magnetotail instability and substorm onset", *J. Geophys. Res.*,92, 13, 456, 1987

Buchner,J., and L.M. Zelenyi, ''Adiabatic and chaotic charged particle motion in curved two-dimensional magnetic field reversals," submitted to *J. Geophys. Res.*, 1988

Buchner, J., and L.M. Zelenyi, ''Regular and chaotic charged particle motion in magnetotail-like field reversals 1. Basic theory of trapped motion," *J. Geophys. Res.*, 94, 11,821, 1989

Chen, J., and P.J. Palmadesso, ''Chaos and nonlinear dynamics of single-particle orbits in magnetotail-like magnetic field,'' *J. Geophys. Res.*, 91, 1499, 1986

Colgate, S.A. (Chair), H. Furth, J.R. Jokipii, C.F. Kennel, L.J. Lanzerotti, E.N. Parker, D. Pines, M. Rosenbluth, and M. Ruderman, *Space Plasma Physics: the Study of Solar System Plasmas,* National Academy of Sciences, Washington, D.C. (1978)

Coppi,B., G. Laval, and R. Pellat, A model for the influence of the earth magnetic tail on geomagnetic phenomena, *Phys. Rev. Letts.*, 16, 1207, 1966

Coroniti, F.V., and C.F. Kennel, Changes in magnetospheric configuration during substorm growth phase, *J. Geophys. Res.*, 77, 3361, 1972a

Coroniti, F.V., and C.F. Kennel, Magnetospheric substorms, *in* Cosmic Plasma Physics, edited by K. Schindler, p. 15, Plenum, New York, 1972b

Coroniti, F.V., and C.F. Kennel, Can the ionosphere regulate magnetospheric convection?, *J. Geophys. Res.*, 78, 2837, 1973

Coroniti, F.V., and A. Eviatar, Magnetic field reconnection in a collisionless plasma, *Astrophys. J. Suppl. Ser.*, 33, 189, 1977

Coroniti, F.V., On the tearing mode in quasi-neutral sheets, *J. Geophys. Res.*, 85, 6719, 1980

Coroniti, F.V., L.A. Frank, D.J. Williams, R.P. Lepping, F.L. Scarf, S. M. Krimigis, and G. Gloeckler, Variability of plasma sheet dynamics, *J. Geophys. Res.*, 85, 2957, 1980

Coroniti, F.V., Space plasma turbulent dissipation: myth or reality? *Space. Sci. Revs.*, 42, 399, 1985a.

Coroniti, F.V., Explosive tail reconnection: the growth and expansion phases of magnetospheric substorms, *J. Geophys. Res.*, 90, 7427, 1985b

Dungey, J.W., Interplanetary magnetic field and the auroral zones, *Phys. Rev. Letts.*, 6, 47, 1961

Evans, D.S., and T.E. Moore, Precipitating electrons associated with the diffuse aurora: evidence for electrons of atmospheric origin in the plasma sheet, *J. Geophys. Res.*, 84, 6451, 1979

Filbert, P.C., and P.J. Kellogg, On the wavelengths of (n = 1/2)fce gyroharmonic emissions in the earth's magnetosphere, *J. Geophys. Res.*, 93, 11, 374, 1988

Fredricks, R.W., Plasma instability at (n + 1/2)fce and its relationship to some satellite observations, *J. Geophys. Res.*, 76, 5344, 1971

Galeev, A.A., and L. M. Zelenyi, Tearing instability in plasma configuration, *Sov. Phys. JETP, Engl. Trans.*, 43, 1113, 1976

Galeev, A.A., F.V. Coroniti, and M. Ashour-Abdalla, Explosive tearing mode reconnection in the magnetospheric tail, *Geophys. Res. Letts.*, 5, 707, 1978

Galeev, A.A., R. Z. Sagdeev, and V. D. Shapiro, *Proc. Joint Varenna-Abastumani Intl. School and Workshop,* (European Space Agency, 1986), p.297, ESA sp.251-ISSN 0379-6566, 1986

Gosling, J.T., J.R. Asbridge, S.J. Bame, W. C. Feldman, G. Paschmann, and N. Sckopke, Solar wind ions accelerated to 40 KeV by shock wave disturbances, *J. Geophys. Res.*, 85, 744, 1980

Gurnett, D.A. R.R. Anderson, F.L. Scarf, R.W. Fredricks, and E.J. Smith, Initial results from the ISEE-1 and -2 plasma wave investigation, *Space Sci. Revs.*, 103, 1979

Hones, E.W., Jr., Plasma flow in the plasma sheet and its relation to substorms, *Radio Sci.*, 8, 879, 1983

Hones, E.W., Jr., Transient phenomena in the magnetotail and their relation to substorms, *Space Sci. Revs.*, 23, 393, 1979

Hones, E.W., Jr., D. N. Baker, S.J. Bame, W.C. Feldman, J.T. Gosling, D.J. McComas, R.D. Zwickl, J.A. Slavin, E.J. Smith, and B.T. Tsurutani, Structure of the magnetotail at 220 R_E and its response to geomagnetic activity, *Geophys. Res. Letts.*, 11, 5, 1984

Huba, J.D., N.T. Gladd, and K. Papadopoulos, Lower hybrid drift turbulence in the distant magnetotail, *J. Geophys. Res.*, 83, 5217, 1978

Huba, J. D., J. F. Drake, and N.T. Gladd, Lower-hybrid-drift instability in field-reversed plasmas, *Phys. Fluids*, 23, 552, 1980

Huba, J.D., N.T. Gladd, and J.F. Drake, The lower hybrid drift instability in nonantiparallel reversed field plasmas, *J. Geophys. Res.*, 87, 1697, 1982

Hubbard, R.F., and T. J. Birmingham, Electrostatic emissions between electron gyroharmonics in the outer magnetosphere, *J. Geophys. Res.*, 83, 4837, 1978

Kennel, C.F., Consequences of a magnetospheric plasma, *Rev. Geophys.*, 7, 379, 1969

Kennel, C. F., and F. E. Engelmann, Velocity space diffusion from plasma turbulence in a magnetic field, *Phys. Fluids*, 9, 2377, 1966

Kennel, C.F., and H. E. Petschek, Limit on stably trapped particle fluxes, *J. Geophys. Res.*, 71, 1, 1966

Kennel, C. F., F.L. Scarf, R.W. Fredricks, J.H. McGehee, and F. V. Coroniti, VLF electric field observations in the magnetosphere, *J. Geophys. Res.*,75, 6136, 1970

Kennel, C.F., and M. Ashour-Abdalla, Electrostatic waves and the strong diffusion of auroral electrons, p. 245 *in* Magnetospheric Plasma Physics, ed. by A. Nishida, D. Reidel, Dordrecht, Holland, 1982

Kennel, C.F., F. V. Coroniti, F. L. Scarf, W. A. Livesey, C. T. Russell, E. J. Smith, K. P. Wenzel, and M. Scholer. "A Test of Lee's Quasi-Linear Theory of Ion Acceleration by Interplanetary Traveling Shocks," *J. Geophys. Res.*, 91, A11, 197, 1986 .

Kennel, C.F., "Cosmic Ray Acceleration: A Plasma Physicist's Perspective", pp. 203-226, *in* "From Particles to Plasmas", Lectures Honoring Marshall N. Rosenbluth (J.W. Van Dam, ed.) Addison-Wesley, N.Y., 1989.

Krymsky, G.F., A regular mechanism for the acceleration of charged particles on the front of a shock wave (in Russian) *Dokl. Akad. Nauk* , SSSR, 243, 1306, (1977)

Kurth, W.S., M. Ashour-Abdalla, L. A. Frank, C. F. Kennel, D. A. Gurnett, D. D. Sentman, and B. G. Burek, "A Comparison of Intense Electrostatic Waves Near fUHR with Linear Instability Theory," *Geophys. Res. Lett.*, 6, 487, 1979

Laval, G.,R. Pellat, and M. Vuillemin, Instabilities electromagnetiques des plasmas sans collisions, *Plasma Phys. Controlled Nucl. Fusion Res., Proc. Intl. Conf. 2nd*, 2, 259, 1966

LeBoeuf, J.N., T. Tajima, C.F. Kennel, and J.M. Dawson, Global simulations of the time-dependent magnetosphere, *Geophys. Res. Letts.*, 5, 609, 1978

LeBoeuf, J.N., T. Tajima, C.F. Kennel, and J. M. Dawson, Global simulations of the three-dimensional magnetosphere, *Geophys. Res. Letts.*, 8, 257, 1981

LeBoeuf, J. N., T. Tajima, and J.M. Dawson, Dynamic magnetic X-points, *Phys. Fluids*, 25, 784, 1982

Lee. M. A., Coupled hydromagnetic wave excitation and ion acceleration upstream of the earth's bow shock, *J. Geophys. Res.*, 87, 5063, (1982)

Lee. M. A., Coupled hydromagnetic wave excitation and ion acceleration upstream at interplanetary traveling shocks, *J. Geophys. Res.*, 88, 6109, (1983)

Lembege, B. and R. Pellat, Stability of a thick two-dimensional quasi-neutral sheet, *Phys. Fluids*, 25, 1995, 1982

Lyon, J.G., S.H. Brecht, J.D. Huba, J.A. Fedder, and P.J. Palmadesso, Computer simulation of a geomagnetic substorm, *Phys. Rev. Letts.*, 46, 1038, 1981

Lyons, L. R., R, M. Thorne, and C. F. Kennel, Electron pitch angle diffusion driven by oblique whistler mode turbulence, *J. Plasma Phys.*,6, 589, 1971

Lyons, L. R., R, M. Thorne, and C. F. Kennel, Pitch angle diffusion of radiation belt electrons within the plasmasphere, *J. Geophys. Res.*, 77, 3455, 1972

Lyons, L.R., and R.M. Thorne, Equilibrium structure of radiation belt electrons, *J. Geophys. Res.*, 78, 2142, 1973

Lyons, L.R., Electron diffusion driven driven by magnetospheric electrostatic waves, *J. Geophys. Res.*, 79, 575, 1974

McIlwain, C.E., Coordinates for mapping the distribution of magnetically trapped particles, *J. Geophys. Res.*, 66, 3681 (1961)

McPherron, R.L., Growth phase of magnetospheric substorms, *J. Geophys. Res.*, 75, 5592, 1970

McPherron, R.L., C.T. Russell, and M.P. Aubry, Satellite studies of the magnetospheric substorms on August 16, 1968,9, phenomenological model for substorms., *J. Geophys. Res.*, 78, 3131, 1973

Meng, C.I., B. Mauk, and C.E. McIlwain, Electron precipitation of evening diffuse aurora and its conjugate electron fluxes near the magnetospheric equator, *J. Geophys. Res.*, 84, 2545, 1979

Ogino, T., A three-dimensional MHD simulation of the interaction of the solar wind with the earth's magnetosphere: the generation of field-aligned currents, *J. Geophys. Res.*, 91, 6791, 1986

Ogino,T. and R. J. Walker, A magnetohydrodynamic simulation of the bifurcation of the tail lobes during intervals with a northward interplanetary magnetic field, *Geophys. Res. Letts.*, 11, 1018, 1984

Ogino, T., R. J. Walker, M. Ashour-Abdalla, and J.M. Dawson, An MHD simulation of By dependent magnetospheric convection and field-aligned currents during northward IMF, *J. Geophys. Res.*, 90, 10835, 1985

Ogino, T., R. J. Walker, M. Ashour-Abdalla, and J.M. Dawson, An MHD simulation of the effects of the interplanetary magnetic field by component on the interaction of the solar wind with the earth's magnetosphere during southward interplanetary magnetic field, *J. Geophys. Res.*, 91, 10029, 1986

Petschek. H.E., Magnetic Field Annihilation, *NASA Spec. Publ., SP-50,* 425, 1964

Petschek, H.E., The mechanism for reconnection of geomagnetic and interplanetary field lines, *in* "The Solar Wind", edited by R. J. Mackin and M. Neugebauer, p. 257, Pergamon, New York, 1966

Petschek, H. E., and C. F. Kennel, Tail flow, auroral precipitation, and ring currents (abstract), *Trans. AGU,* 47, 137, 1966

Roeder, J.L., and H.C. Koons, A survey of electron cyclotron waves in the magnetosphere and the diffuse auroral precipitation, *J. Geophys. Res.*, 94, 2529, 1989

Russell, C.T., and R.L. McPherron, The magnetotail and substorms, *Space. Sci. Rev.,* 15, 205, 1973

Scarf, F.L., R.W. Fredricks, C.F. Kennel, and F.V. Coroniti, Satellite studies of magnetospheric substorms on August 15, 1968, *J. Geophys. Res.*, 78, 3119, 1973

Schindler, K., A theory of the substorm mechanism, *J. Geophys. Res.*, 79, 2803, 1974

Schumaker, T. L., M. S. Gussenhoven, D.A. Hardy, and R. L. Carovillano, The relationship between diffuse auroral and plasma sheet electron distributions near local midnight, *J. Geophys. Res.*, 94, 10,061, 1989

Sharber, J.R., The continuous (diffuse) aurora and auroral-E ionization, *in* Physics of Space Plasma, ed. by T.S. Chang, B. Coppi, and J.S. Jasperse, SPI Conference Proceedings and Reprint Series, vol. 4, Scientific Publishers, Cambridge, Mass., 1981

Shaw, R.R., and D.A. Gurnett, Electrostatic noise bands associated with the electron gyrofrequency and plasma frequency in the outer magnetosphere, *J. Geophys. Res.*, 80, 4259, 1975

Solomon, J., N. Cornilleau-Wehrlin, A. Korth, and G. Kremser, Generation of ELF electromagnetic waves and diffusion of energetic particles in steady and non-steady situations in the earth's magnetosphere, p. 119 *in* "Plasma waves and Instabilities at Comets and in

Magnetospheres", (ed. by B.T. Tsurutani and H. Oya) Geophysical Monograph 53, American Geophysical Union, Washington, D.C. 1989

Tan, L.C., G.M. Mason, G. Gloeckler, and F.M. Ipavich, Energetic particle diffusion coefficients upstream of quasi-parallel interplanetary shocks, *J. Geophys. Res.*, 94, 6552, 1989

Teresawa, T., Numerical study of explosive tearing mode instability in one-component plasmas, *J. Geophys. Res.*, 86, 9007, 1981

Vasyliunas, V. M., independent research described in C. F. Kennel, *Rev. Geophys.*, 7, 379, 1969

Walker, R. J., and T. Ogino, Field-aligned currents and magnetospheric convection: a comparison between MHD simulations and observations, *in* "Modeling Magnetospheric Plasma" (Geophys. Monograph, ser. 44), T. E. Moore and J.H. Waite, Jr., eds., p. 39, AGU, Washington, DC, 1988

Walker, R. J., and T. Ogino, Global magnetohydrodynamics of the magnetosphere, *IEEE Trans. Plasma Sci.*, 17, 135, 1989

Winningham, J.D., F. Yasuhara, S.-I. Akasofu, and W.J. Heikkila, The latitudinal morphology of 10 eV- 10 keV electron fluxes during magnetically quiet and disturbed times in the 2100-0300 MLT sector, *J. Geophys. Res.*, 80, 3148, 1975

Young, T.S.T., J.D. Callen, and J.E. McCune, High frequency electrostatic waves in the magnetosphere, *J. Geophys. Res.*, 78, 1082, 1973

Zelenyi, L.M., A.A. Galeev, and C.F. Kennel, "Ion Precipitation from the Inner Plasma Sheet due to Stochastic Diffusion", accepted, *J. Geophys. Res.*, 1989

ROLE OF SIMULATIONS IN FUTURE MAGNETOSPHERIC PROGRAMS

Maha Ashour-Abdalla[1,2] and Ferdinand V. Coroniti[1,3]

[1]Department of Physics
[2]Institute of Geophysics and Planetary Physics
[3]Department of Astronomy
University of California, Los Angeles
Los Angeles, California 90024 U.S.A.

ABSTRACT

In this paper an attempt is made to assess the impact of simulations on space plasma research. Examples where simulations have been a successful complement to observations and analytical theory are presented. Some current research where simulation techniques are stressed up to and beyond their capabilities are discussed. The major challenges we face in advancing simulation techniques are outlined. Finally, it is argued that even with current techniques, "mission oriented" simulations can have a major influence on upcoming future missions.

1. INTRODUCTION

During the last decade simulation techniques have become an integral part of space plasma research. In this paper, we will attempt to assess the impact of simulations on space physics and suggest possible future directions for simulation research.

We first take a retrospective view on simulation (Section 2) and discuss very briefly some examples where simulations have been a successful complement to observations and analytical theory. We then discuss some current research where simulation techniques are stressed up to and beyond their capabilities (Section 3). Finally (Section 4) we show that even with current techniques, simulation can have a major influence on upcoming future missions.

2. A RETROSPECTIVE VIEW OF SPACE SIMULATIONS

With the advent of the NASA Solar Terrestrial Theory Program, simulations became an integral part of space plasma research. For the purpose of assessing the impact of simulations, we will define a successful simulation as one which either explains observations or advances our understanding of fundamental theory. There are many examples of such successes, but in this paper we will discuss only two of each category. Since the papers reporting these studies have already been published, we will keep our discussion brief and refer the interested reader to the original papers for a more detailed description.

The first example where simulations have helped to explain observations is the acceleration of ionospheric ions on auroral field lines and the formation of conical distributions. Figure 1 shows a schematic of the auroral field lines and summarizes our present understanding of the phenomenology. In the upper part are the well known double layers and associated parallel electric fields which produce upward going ion beams and the downward auroral electron beam. The accelerated electrons carry an upward field-aligned current which is closed through the ionosphere and the magnetosphere by a downward directed current carried by cold electrons. The observed intensity of these field aligned currents [Mozer et al., 1980] is large enough to destabilize ion cyclotron waves [Kindel and Kennel, 1971]. The problem was to explain how ion cyclotron waves could accelerate ionospheric ions to high energies and produce the observed conic distribution functions.

The early work on ion cyclotron waves by Drummond and Rosenbluth [1962] showed that the waves would saturate at a low amplitude $e\phi/T_e << 1$, due to the formation of an electron plateau. This prediction was confirmed by early simulations [Pritchett et al., 1981] which indicated that the ions were heated by less than a factor of 2. Clearly, if ion acceleration was to be efficient, electron plateau formation had to be suppressed, thus allowing the waves to grow to higher amplitudes. Kindel [1970] had already suggested that finite interaction length effects might remove the electron plateau. In order to model a finite length auroral field line, Ashour-Abdalla and Okuda [1983] developed a simulation (constant flux model) in which the drifting electrons only enter at the ionospheric end of the field line which is treated as a boundary condition for the simulation model. The electron distribution is prescribed as a drifting Maxwellian at $x = 0$, and the distribution develops a plateau when it leaves the system at $x = l$ as a result of self-consistent wave-particle interactions (Figure 2). These simulations resulted in wave growth to large amplitude. Figure 3 is a plot of the ion distribution $f(v_\perp^2)$ versus $v_\perp^2/v_{ti}^2$ where $v_\perp$ and v_{ti} are the velocity component perpendicular to the magnetic field and the ion

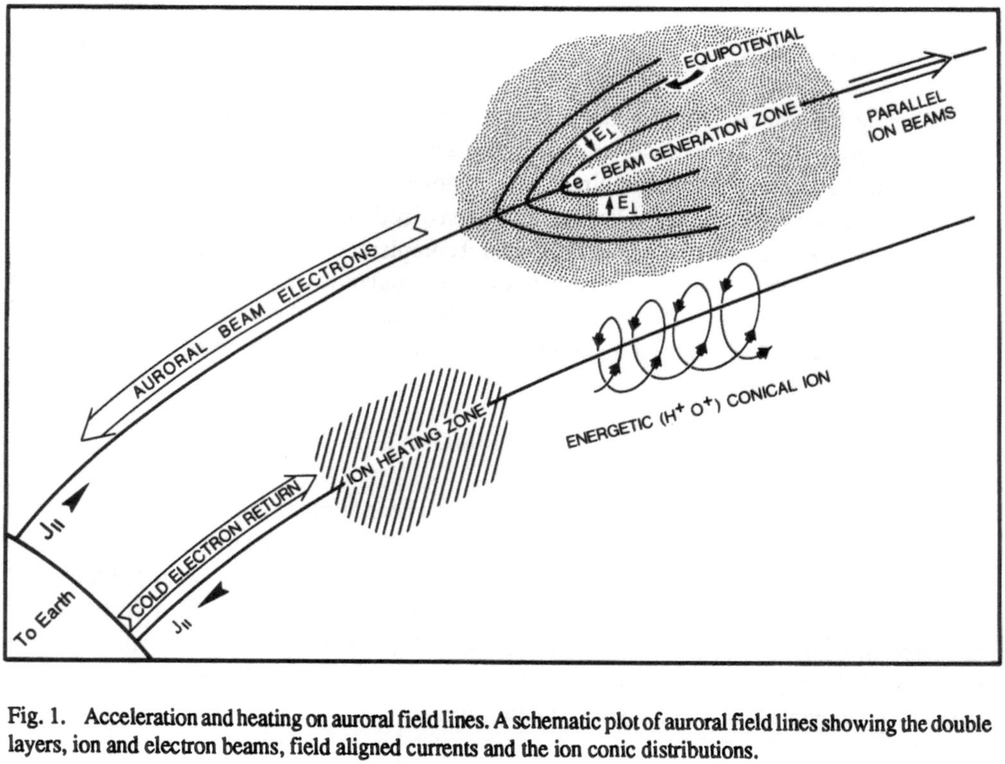

Fig. 1. Acceleration and heating on auroral field lines. A schematic plot of auroral field lines showing the double layers, ion and electron beams, field aligned currents and the ion conic distributions.

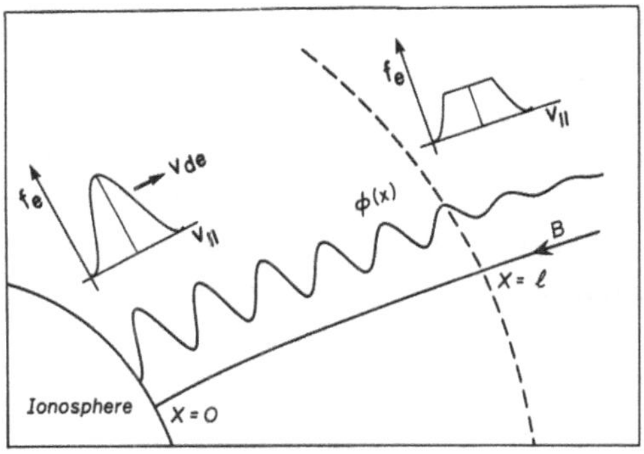

Fig. 2. A model for auroral field lines in which the ionosphere is the source of cold drifting Maxwellian electrons streaming upwards. As the electrons move up, ion cyclotron turbulence develops, causing a plateau on the electron velocity distribution and ion perpendicular heating. The free energy source of ionospheric electrons is continuously replenished [from Ashour-Abdalla and Okuda, 1983].

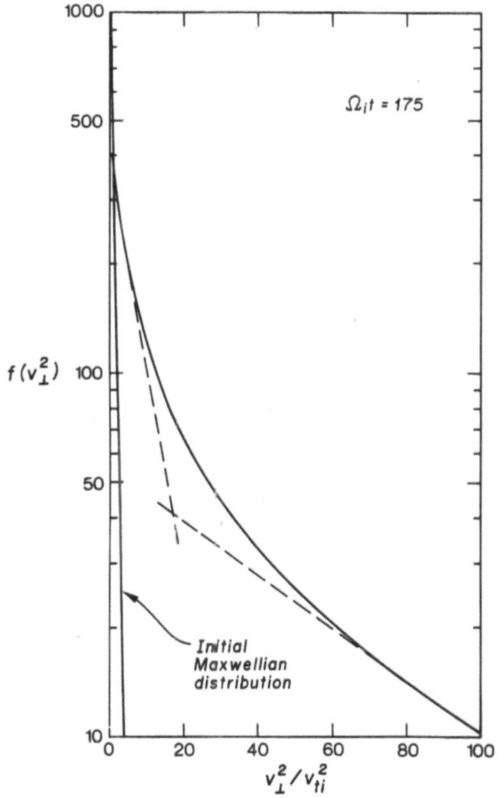

Fig. 3. The ion velocity distribution at low altitudes as a function of energy at $\Omega_i t = 175$. Note the distribution may be divided into bulk and tail distributions in which the tail temperature may be as large as 100 times the initial temperature [from Ashour-Abdalla and Okuda, 1983].

thermal velocity respectively; the initial Maxwellian distribution is shown for comparison. The final distribution function is divided into two parts, a bulk distribution and a high energy tail. The temperature of the high energy tail is about 50 to 100 times the initial temperature. Since the parallel heating remains small, the final distribution resembles a conic distribution.

A second example where simulations have been useful in explaining observations is related to the dynamics of the magnetotail during magnetospheric substorms and the formation of plasmoids. In the phenomenological substorm model, a new near earth neutral line forms at breakup [Hones, 1979] with the subsequent ejection (tailward) of a plasmoid when the last closed field line is reconnected. Using observations from ISEE-3, Hones et al. [1984] confirmed the plasmoid ejection. Birn [1984] used a 3-D magnetohydrodynamic simulation to show that plasmoids could self-consistently form via a tearing instability in a tail-like magnetic field configuration (Figure 4). Plasmoid ejection was caused by the earthward pressure gradient due to the flaring of the tail. The ejection speed was found to be consistent with observations.

The first example in which simulations have advanced fundamental theory concerns the reflected ion dissipation in fast shocks. Early theoretical models of ion dissipation by Sagdeev [1966] and Auer et al. [1971], involved the reflection of some fraction of the upstream ions by the magnetic field and electrostatic potential in the shock ramp. After reversing ($v_x < 0$) velocity, the reflected ions are accelerated by a motional electric field E_y, and gain energy. Eventually

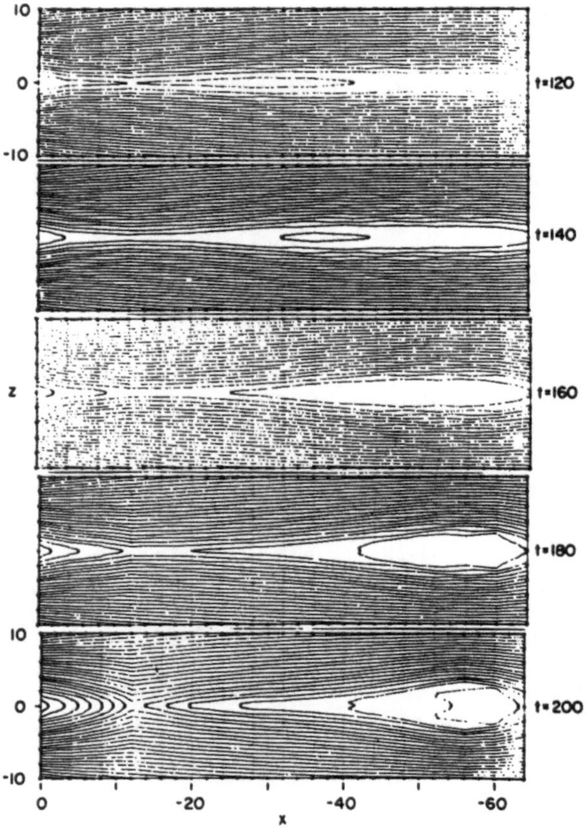

Fig. 4. Magnetic field lines in the noon-midnight meridian from a simulation of reconnection in the magnetotail. The time sequence shows the evolution of a plasmoid in the tail [from Birn, 1984].

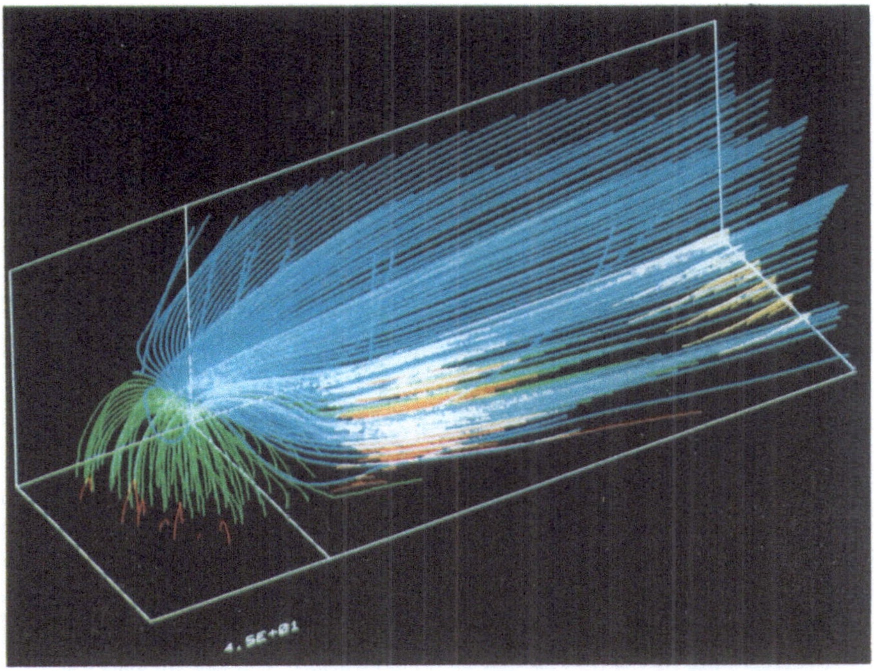

Fig. 6. Magnetic field lines from a global magnetohydrodynamic (MHD) simulation of the interaction of the solar wind with the magnetosphere. In this case, the interplanetary magnetic field was southward. Green magnetic field lines are closed, blue field lines are open while the red and yellow field lines are detached from the Earth. [from Walker and Ogino, 1989].

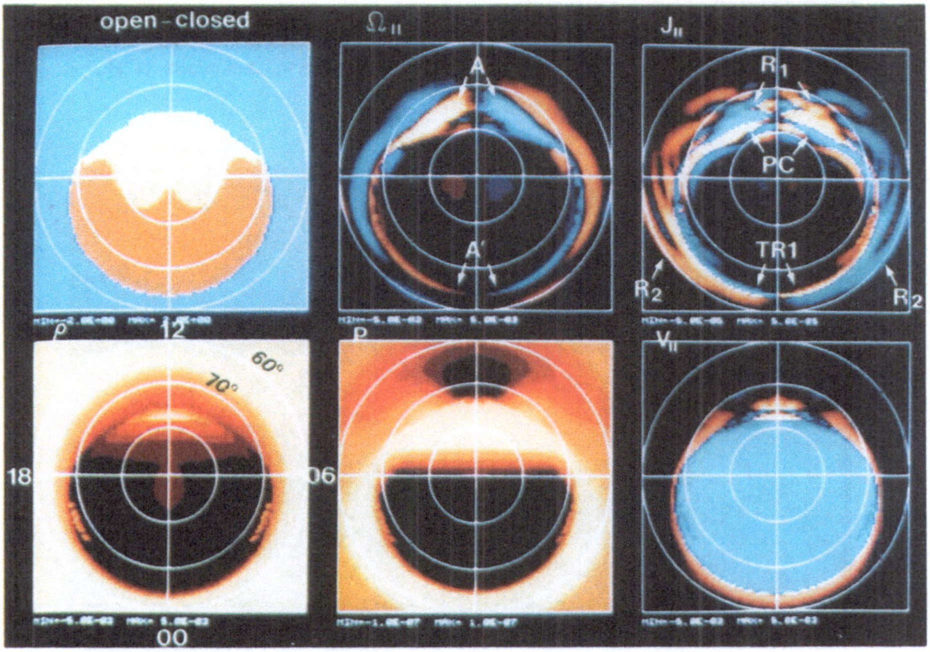

Fig. 7. Polar projections of the regions of open and closed field lines, the parallel vorticity ($\Omega_{\parallel}$), the parallel current density ($J_{\parallel}$), the density (ρ), the pressure (P) and the parallel velocity ($v_{\parallel}$) for the case with global MHD simulation, southward IMF. The open field line region is orange. For the parallel vorticity, the parallel current and the parallel velocity values parallel to $\underline{B}$ are red and yellow, and values antiparallel are in blue. Yellow and light blue represent larger amplitudes than red or dark blue. For the density and pressure, orange and yellow represent the largest values while dark red and black represent small values [from Walker and Ogino, 1989].

Plate 1

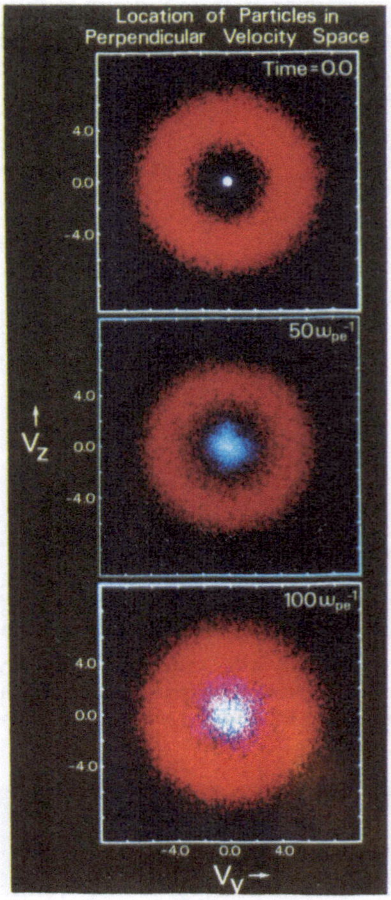

Fig. 11. Cold electron heating observed in a numerical simulation of 3/2 instabilities. The top inset shows the initial ring distribution of hole electrons, colored red, whose free energy caused a strong 3/2 instability. The cold electrons are blue. By t = 50ωp^{-1}, the instabilities saturated; at this point, the cold electrons were heating rapidly. As the wave amplitudes died out, cold electrons continued to heat and gradually lost their identity. The saturation amplitude was several hundred μV/m scaled to the magnetosphere [from Ashour-Abdalla et al., 1980].

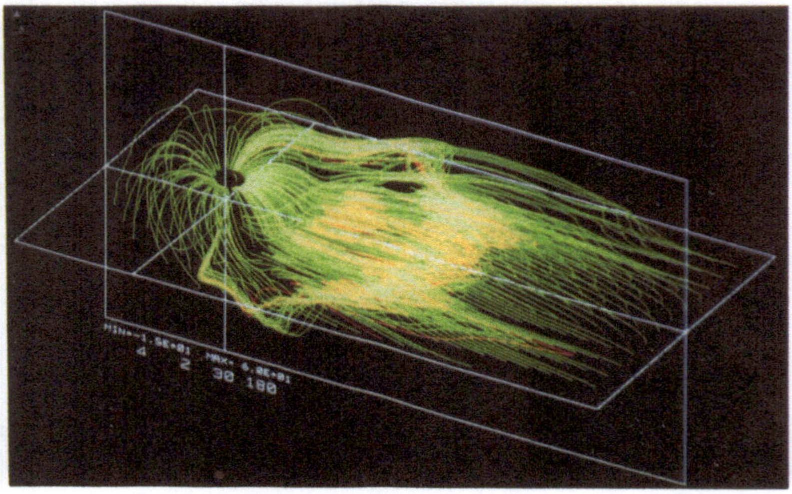

Fig. 14. Magnetic field lines from a global MHD simulation for the case when the IMF had both dawnward and southward components. Green field lines are closed while yellow field lines start at the equator and return to the equator before they reach the Earth. The plasmoid field lines start in the southern dusk polar region, spiral about the plasmoid and return to the northern dawn polar region [after Ogino et al., 1989].

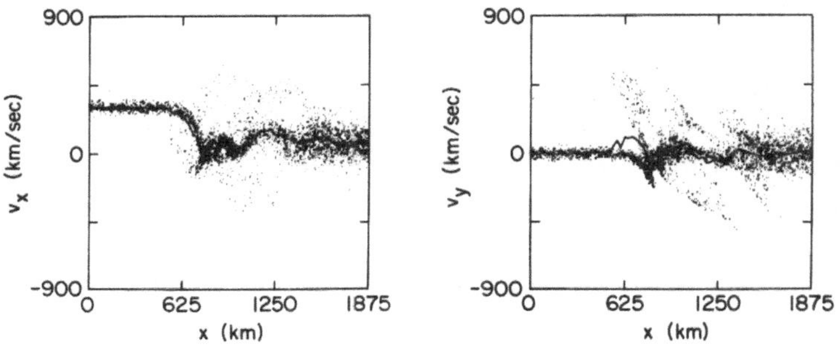

Fig. 5. A phase space plot at t = 8 s. The left hand panel shows vx versus x phase space. The right hand panel shows vy versus x phase space. Solid lines denote the mean ion velocities V_x and V_y [from Leroy et al., 1981].

they penetrate downstream due to cyclotron motion. The central question was: can a self-consistent quasi perpendicular shock be constructed based on ion reflection in which the magnetic field and potential adjust to reflect the correct number of ions to satisfy the Rankine-Hugoniot relations? Leroy et al. [1981] carried out a hybrid simulation (kinetic ions-fluid electrons), and showed that the reflection process does form a steady state high Mach number shock (Figure 5).

Our last example concerns the large scale structure of the magnetosphere. The early work of Axford and Hines [1961], Dungey [1961] and Levy et al. [1964], assumed that large scale magnetospheric dynamics could be described by magnetohydrodynamics. However, it was not obvious that MHD would actually apply to the highly collisionless space plasmas or that a self-consistent MHD solution would actually produce the popular cartoons derived from MHD reasoning (the pioneers in this field never actually solved the equations, but only discussed them).

Results from a numerical simulation of the three-dimensional MHD flow of the solar wind around the magnetosphere are shown in Figure 6. This figure is a snapshot of the magnetic field configuration for southward IMF which clearly corresponds to the reconnected magnetic topology inferred from MHD reasoning. The simulation also confirms the convection pattern drawn by earlier MHD theorists such as Axford and Hines [1961] and Brice [1967]. Figure 7 shows the field aligned currents in the simulation which develop as a consequence of coupling the MHD stresses in the magnetosphere to the ionosphere. Thus the simulation also reproduces the observed Region 1 and 2 current systems. Having confirmed the MHD cartoons, simulations can now be used to determine magnetospheric configuration that are not so obvious at the cartoon level.

3. FUTURE SIMULATION CHALLENGES

The above examples illustrate past successful applications of simulations to space plasmas. In the future, however, we face some severe challenges as we try to make simulations more relevant to observations. In this section we discuss three examples in our recent experience where the present limitations of simulation techniques constrain our ability to model space plasma behavior. As an example of kinetic instabilities we first present a recent attempt to understand weak instabilities from gentle free energy sources. As an example of simulations which reside between kinetic and MHD, we discuss the progress on the study of reconnection in the Earth's magnetotail. Finally, we will discuss the limitations of the present global magnetohydrodynamic simulations in modeling the magnetosphere.

Electrostatic wave emissions with frequencies between the ion and electron plasma frequencies have been observed in many different space environments [Rodriguez and Gurnett, 1975; Gurnett et al., 1976; Kennel et al., 1982; Scarf et al., 1984 a,b]. These waves have several properties in common: 1) they have weak amplitudes $E^2/8\pi\, nkT_e \sim 10^{-5} - 10^{-6}$; 2) they occur over large regions in space; and 3) in a standard plasma there are no normal modes in this frequency range. However there can be special beam modes or special mixtures of hot and cold plasma. For example, these special mixtures of hot and cold plasma have been very successful in explaining the near Earth broadband electrostatic noise (BEN) [Grabbe and Eastman, 1984; Dusenbery and Lyons, 1985; Ashour-Abdalla and Okuda, 1986a,b; Schriver and Ashour-Abdalla, 1987].

Recently, however, Coroniti and Ashour-Abdalla [1989] argued that a class of narrow-band electrostatic noise (NEN), which is observed in association with slow shocks in the distant tail, might not be explicable as a conventional instability. They suggested that NEN might represent a non-standard plasma wave mode which results from a hole in the low energy electron distribution. They proposed a simple model of how the hole might form due to the interaction of the tail electrons with the standing pair of slow shocks (see Figure 8). The hole supports a non-standard electrostatic wave – the hole mode – which has a low parallel phase speed and propagates nearly along the magnetic field. The hole can also produce a positive slope

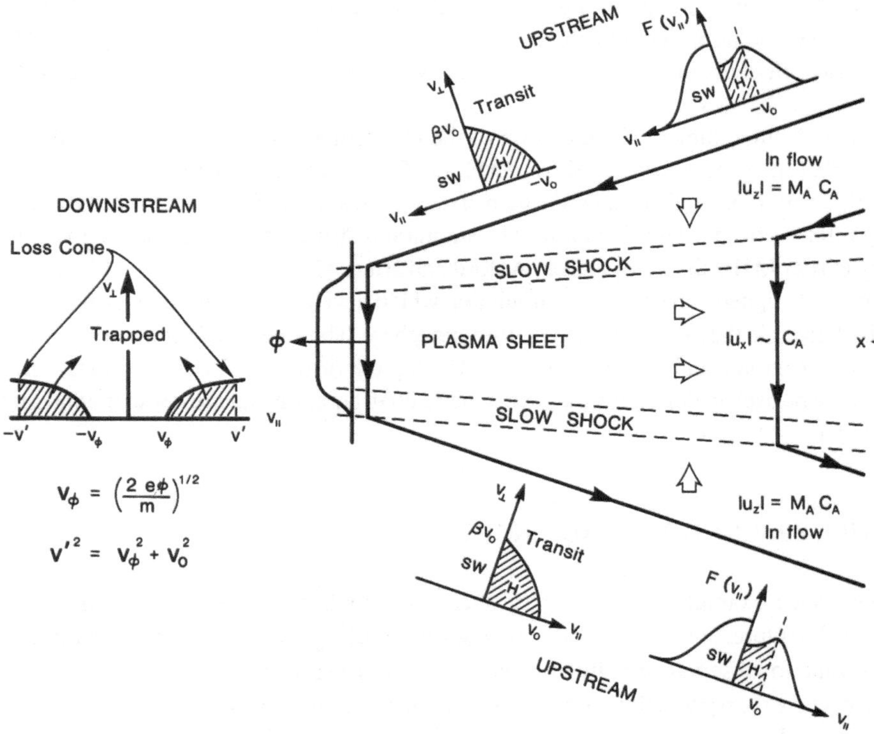

Fig. 8. A sketch of the slow shock structure, the shock electrostatic potential ϕ, the $v_\perp - v_\parallel$ upstream and downstream electron velocity space, and the electron distribution integrated over $v_\perp(Fe(v_\parallel))$. SW indicates electrons which come directly from the solar wind. H indicates the velocity space hole. Transit indicates electrons that pass through the plasma sheet to the opposite tail lobe [from Ashour-Abdalla et al., 1989].

in the parallel electron distribution (after integrating over $v_\perp$) which destabilizes the mode. The wave growth rate is quite small ($\gamma/\omega_{pe} \sim 10^{-3}$), but does cover the observed NEN frequency range ($0.01 < \omega/\omega_{pe} < 0.1$).

A fundamental issue is whether such fine scale velocity space structures really can persist in the plasma or whether they will be rapidly destroyed by their associated instabilities. To answer this question, the obvious next step was to consider a particle simulation. Ashour-Abdalla et al. [1989], used a one-dimensional electrostatic particle code to simulate the hole mode. Since the instability has a weak growth rate and low saturation amplitude, we found that the usual random loading in velocity space of the typical simulation with $\sim 10^5$ particles is inadequate. Numerical collisions, which are particularly strong due to the steep gradient in the electron distribution, rapidly destroyed the hole on time scales faster than the growth time. Thus to model the collisionless evolution of the hole it was necessary to use a quiet start [Denavit and Walsh, 1981]. However the quiet start configuration has an instability in which neighboring sets of particles in phase space interact in the same manner as a large number of small beams. In order to decrease the effect of the multi-beam instability and clearly observe the growth it was necessary to use ~1.6 million particles.

Figure 9 shows a plot of electron phase space ($v_\perp$ - $v_\parallel$) at different times during the course of the simulation. At t = 0 (Figure 9a) the initial hole is well defined and has a significant region

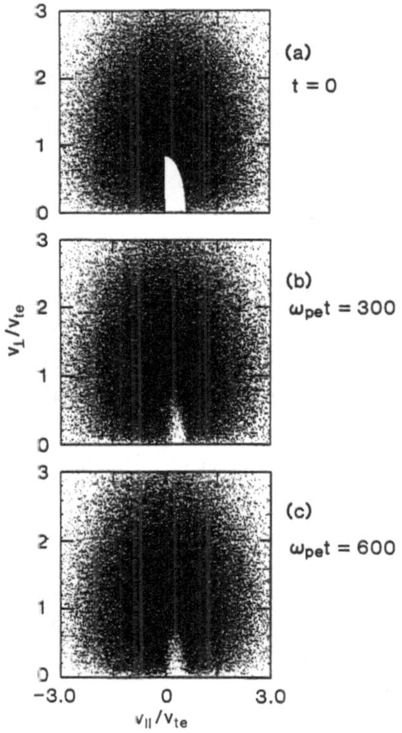

Fig. 9. A plot of phase space ($v_\parallel$ - $v_\perp$) at different times during the simulation run. Panel (a) corresponds to the initial distribution and clearly shows the presence of a hole. Panel (b) is taken at time $\omega_{pe}t = 300$ when the hole is starting to fill in. Panel (c) is at $\omega_{pe}t = 600$ and shows that the slope of the electron distribution is weakened but the hole persists [from Ashour-Abdalla et al., 1989].

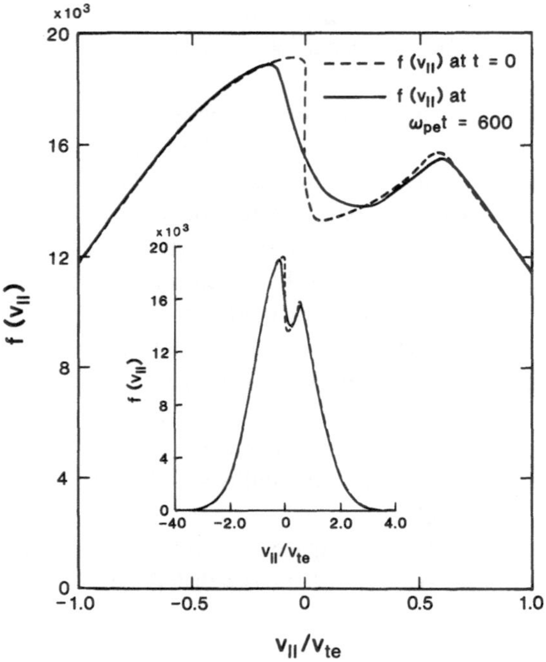

Fig. 10. The initial (dotted lines) and the final (solid lines) parallel electron distributions, $F_e(v_\parallel)$, are plotted versus $v_\parallel/v_{te}$ on an expanded scale. The right hand insert shows the complete electron distribution [from Ashour-Abdalla et al., 1989].

of positive slope. At $\omega_{pe}t = 300$ (Figure 9b) the hole starts filling in and its boundaries become diffuse; $\omega_{pe}t = 300$ corresponds to the time of peak amplitude. In runs with fewer particles, (~ 10^5), and identical initial conditions the hole was completely filled by $\omega_{pe}t = 50$ due to numerical collisions; however, for stable runs with a large number of particles ~1.6 million, the hole persisted with only slight filling. At $\omega_{pe}t = 600$ (Figure 9c) the slope of the electron distribution is further reduced and the instability ceases. Figure 10 shows the initial (dotted lines) and the final parallel distribution, $f(v_\parallel)$, plotted versus $v_\parallel$. The observed flattening of $f(v_\parallel)$ results from collisionless wave-particle trapping and not numerical diffusion. Thus the hole mode remains as an eigenmode in the saturated state. This simulation is one of the few that has been able to reproduce these weak instabilities.

At this point, it is interesting to go back in history and compare these NEN simulations with one of the first simulation attempts to explain space plasma waves, the electron cyclotron harmonic waves at $(n + 1/2)\Omega_{ce}$ [Ashour-Abdalla et al., 1980]. In this simulation computational constraints required a steep free energy source with an effective temperature anisotropy A ~ 1. Naturally this large free energy led to fast growth rates and huge saturation amplitude. Moreover, as can be seen in Figure 11 the large amplitude waves caused violent trapping and heating of the cold plasma. The observations by Kurth et al. [1980] and Rönnmark et al. [1978] show however, that the electron velocity distribution function at best has a gentle slope. Also, the wave amplitudes are ~ 100 μV/m [Belmont et al., 1983; Roeder and Koons, 1989] and are much weaker than had been reported by OGO-5 [Kennel et al., 1970].

The obvious step to model the $(n + 1/2)\Omega_{ce}$ observations more realistically would be to apply the same methodology as we did for NEN, that is, a quiet start and a large number of particles. Unfortunately the physics of the odd-half harmonic waves is undoubtedly more complicated than can be represented in a simple homogeneous plasma simulation. The observations show the strongest waves to be confined to the equator [Kennel et al., 1970; Fredricks and Scarf, 1973], but the particles are bouncing back and forth along the field lines. So wave propagation calculations in an inhomogeneous field are required. Furthermore the replacement of the precipitating hot electrons by cold electrons (which are in the loss cone) must be included since charge neutrality must be obeyed. Hence the odd half-harmonic waves represent a difficult problem to simulate because, although the waves are spatially localized, the physics of these waves involves the large scale dynamics. Thus the challenge here is to develop kinetic simulations that incorporate the essential large scale interactions.

Although simulations of small scale kinetics have their difficulties, there is another scale where simulations should have an impact – a scale between kinetic theory and magnetohydrodynamics – the most prominent example is reconnection.

3.2. Kinetic Reconnection in the Magnetotail

A fundamental problem in reconnection theory is to explain the formation of the substorm neutral line within the closed field line region of the plasma sheet. Recently Pritchett et al. [1989] investigated the ion tearing growth using a 2-D kinetic equilibrium model for a self-consistent flaring tail [Lembege and Pellat, 1982]. In these simulations the response of a current sheet near marginal stability (Figure 12) to an external, low-frequency perturbation is studied. Figure 13 shows results when the external perturbation serves as a pump which can drive the modes of the system. After a time $\Omega_0 t = 64$, the modulation of the Bz field produces an extended neutral sheet region with nearly parallel field lines. In this region the particles are no longer magnetized, and this restores the Landau damping which had been removed by the initial presence of the finite normal field. The time evolution to a new equilibrium containing magnetic islands is shown in the lower panel (Figure 13).

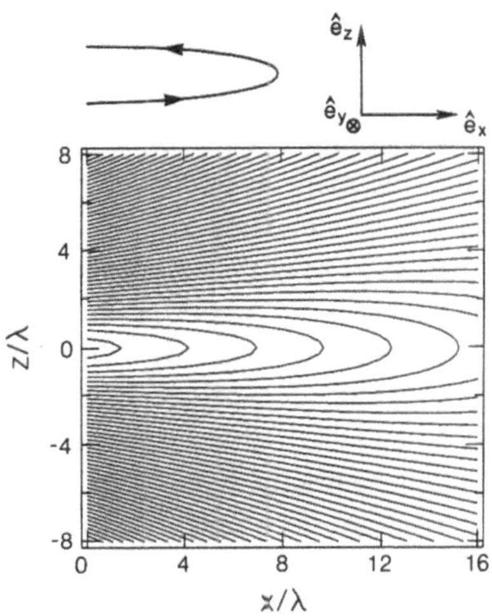

Fig. 12. Coordinate system and magnetic field lines for the quasi-neutral sheet equilibrium [from Pritchett et al., 1989].

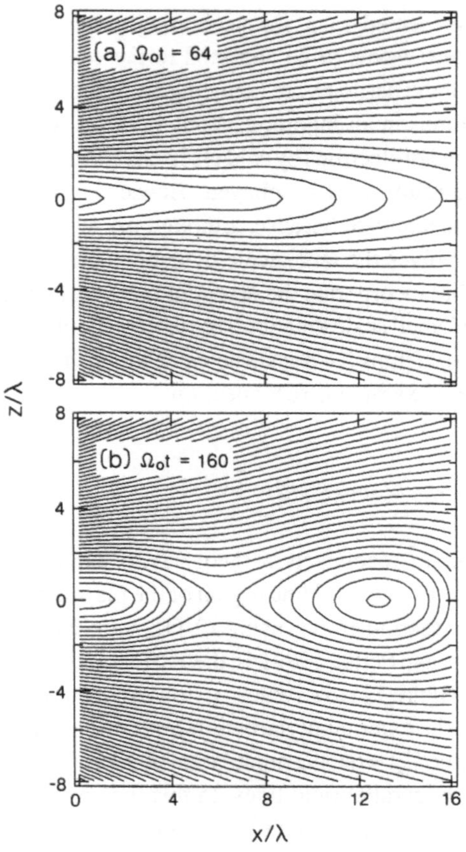

Fig. 13. Magnetic field lines in the driven simulation at (a) $\Omega_o t = 64$ and (b) $\Omega_o t = 160$ [from Pritchett et al., 1989].

Although these simulations could help to explain how x-type neutral lines are formed in the near-earth plasma sheet, their applicability is quite limited. First, only one species is considered in the simulation thus the important dynamics of the electrons are neglected. Second, the parameter regime which can be simulated today is limited to very thin current sheets. Although the equilibrium in Figure 12 resembles a tail-like configuration, the plasma parameters are far from tail-like. The thickness of the current sheet layer (λ) is of the order of the ion Larmor radius based on the strong external magnetic field. Finally, current theoretical models of the tearing mode emphasize the behavior of particle orbits, such that stochastic orbits can lead to instability [Büchner and Zelenyi, 1986]. However, in simulations, particle orbits are interrupted by hitting walls. Since charge neutrality requires that particles be returned to the box, the reflection or return of particle introduces orbital dynamics that do not represent the conditions in space where particles can leave a region and not necessarily return.

Thus the challenge in this mid-size scale is to develop global kinetic codes that can adequately model the particle dynamics for physical parameters of the tail. Naturally, since reconnection is inherently a three-dimensional problem, the ultimate challenge is to study reconnection in a fully 3-D system.

3.3. *Flux Ropes in the Tail*

Over the past several months Ogino and colleagues have been trying to model magnetospheric dynamics during substorms [Walker et al., 1988; Walker and Ogino, 1989; Ogino et al., 1989]. In these studies they have used an MHD model to simulate the changes in the magnetospheric configuration which occur following a southward turning of the IMF. These calculations have reproduced many of the features of the near-Earth neutral line model of substorms. In the simulations, dayside reconnection is followed by reconnection in the nightside plasma sheet. The nightside reconnection starts first on closed field lines within the plasma sheet, then when lobe field lines start to reconnect, a plasmoid is released down the tail.

Recently, they have examined the case when the IMF had an east-west component as well as a southward component. The overall changes in the magnetospheric configuration are the same in this case as in the case with just a southward IMF. However the configuration of the plasmoid is different. For a purely southward IMF the plasmoid is a magnetic O region threaded by field lines which close on themselves in the shape of the letter O. Plotted in Figure 14 is the complex magnetic geometry which results when the IMF has a B_y component. In this case the plasmoid consists of a twisted flux rope which extends from the northern dawn polar region through the center of the tail and is connected to the Earth in the southern dusk polar region. In the center of the tail the plasmoid is draped with open field lines with both ends in the solar wind. As the plasmoid evolves, first some, then all of the field lines which enter the poles in Figure 14 become attached to the IMF and the plasmoid escapes the Earth's magnetosphere.

Here the problem is how did B_y get into the tail in the first place? The answer is that the present simulations are so resistive that the solar wind B_y simply diffuses into the tail lobes to produce a tail B_y around which the plasmoid can form. However the magnetosphere does not allow rapid diffusion of the solar wind field. Thus the challenge is to develop less dissipative MHD codes and to increase the realism of the MHD dynamics.

3.4. *Discussion*

In conclusion, there are challenges ahead of us in all different scales of simulations. A new generation of simulations will need to be developed so that space plasma regimes can be realistically modeled. Clearly, just like developing an instrument, it will be a continuous struggle to improve upon the use of simulations. Factors that will help us are:

1) bigger and faster computers,
2) better graphic displays,
3) new codes which model part of the physics so that other parts can be done at higher resolution.

Both new computers and new codes will come along on a long and uncertain time scale. However, before their advent, a new round of space observations – the ISTP program – will arrive. Even the present generation of simulation techniques, if we further refine them, can make an important contribution to the next era of space plasma research – the ISTP era.

4. MISSION ORIENTED SIMULATIONS

At present the comparison of simulations and observations is particularly difficult since the two outputs are cast in completely different forms. A standard simulation of wave-particle interactions shows a time history of the wave energy whereas the experimentalists will show

Sweep Frequency Receiver plots. For particles, simulators show phase space, which cannot be measured by experimentalists. Clearly we can do better by attempting to unify the data output from both simulations and observations.

During ISTP simulation data should be cast in data format which can be analyzed and processed through the same software which is used to display satellite measurements. In effect we should create virtual spacecraft within the simulation box whose local measurements create simulated data time-series. The simulated data can be processed through the same algorithms used for spacecraft measurements (Figure 15). This output will be particularly useful for spatially inhomogeneous simulations where virtual spacecraft located in different regions will "observe" different local dynamics of some larger scale plasma processes; thus simulated time series from a cluster of virtual spacecraft could help in the interpretation of local spacecraft

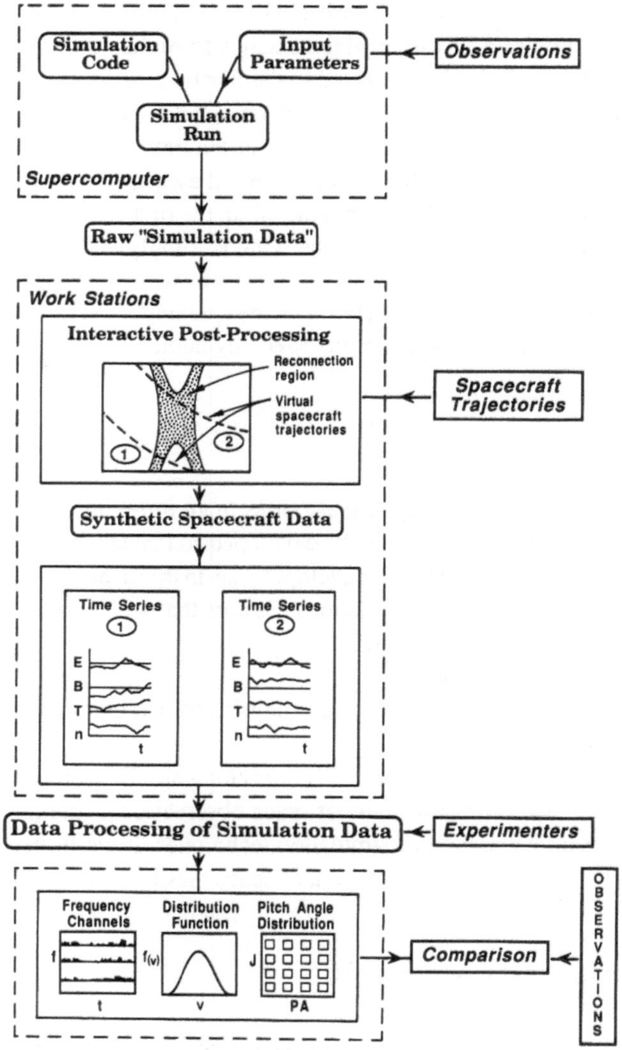

Fig. 15. ISTP simulation data analysis. A flowchart of how simulation data should be processed in order to facilitate comparison with data.

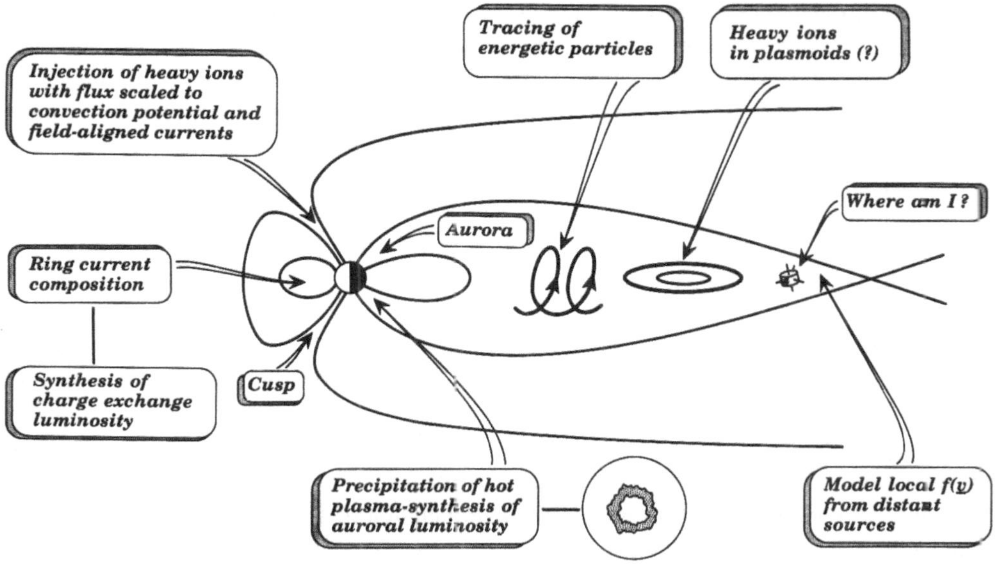

Fig. 16. Global models for ISTP. A schematic showing different studies using global models of the Earth's magnetosphere.

observations in terms of more global behavior. This step is a key to the development of mission oriented simulations which would render simulations more useful in the interpretation of spacecraft data.

We have by no means exhausted the potential of present global MHD codes to provide important diagnostic information about the magnetosphere. Here the emphasis will be to incorporate phenomenology acquired from spacecraft observations into our present codes (see Figure 16). For example ionospheric heavy ions from the cusp and auroral regions are injected into the magnetospheric internal convection flow. Therefore an ionospheric source could be added to the present global models by placing a heavy ion emitter at the ionosphere. The injected flux of heavy ions could be scaled to the polar cap potential and the field aligned current strength, which are both computed in the global codes. Since the heavy ions follow convection, the simulated ions can trace the heavy ion composition gradient in the lobe of the tail, the injection of heavy ions into the ring current, their penetration of the dayside magnetopause, and their loss down the tail and into the magnetosheath. In the ring current region the global model could be used to construct 2-D images of charge exchanged atoms. Charge exchange leaves the oxygen and helium ions in excited states which radioactively decay. By adding charge exchange to the global models, the integrated line of sight luminosity could be predicted and compared with future optical images of the ring current.

The global models could also provide information on the changes in the flux of energetic particles in the magnetosphere by tracing particle orbits in the model's self-consistent electric and magnetic fields. Understanding the acceleration of energetic particles can be achieved by comparing the calculated fluxes with local observation. The models can also give important information on the precipitation flux on auroral field lines. In the limit of strong diffusion the precipitated flux equals the trapped flux. Since the global models determine the pressure on every field line, these codes can output the precipitation energy flux hitting the atmosphere. Going one step further, a phenomenological model that relates the auroral luminosity to the precipitation energy flux could be used to create a simulated auroral oval which can be compared with polar images.

One of the hardest problems found by researchers studying the deep tail observations from ISEE-3 was simply determining where the spacecraft was in relation to the magnetospheric boundaries. Certainly the global models can be used to alleviate this difficulty. For example, by using upstream solar wind parameters in the global model and calculating the time history response of the magnetosphere, one could determine where the spacecraft is relative to magnetospheric boundaries.

5. SUMMARY

In summary, the status of simulations in space research is secure, and simulations promise to make a significant contribution to future magnetospheric missions. Simulations have proven successful for advancing fundamental theory and in explaining observations. We face many challenges in improving numerical techniques for both kinetic and magnetohydrodynamic simulations. However, as in the case of NEN, when simulation techniques are pushed hard, they provide important insights in the interpretation of space observations. Although simulation technology will certainly advance, even the present codes and simulation techniques offer many exciting and important studies that can have a major impact on the future upcoming missions.

ACKNOWLEDGEMENTS

We would like to thank Jean Berchem, Margaret Chen, Rick Richard and Ray Walker for their useful comments and discussions. We would also like to thank Tomik Ebrahimian for help with the computing. This research was supported by NASA Solar-Terrestrial Theory Program grant NAGW-78 and ISTP grant NAG5-1100. Computing was performed on the Scientific Computer Systems SCS-40 at UCLA and on the Cray X-MP at the San Diego Supercomputer Center (SDSC).

REFERENCES

Ashour-Abdalla, M., J.N. Leboeuf, J.M. Dawson, and C.F. Kennel, A simulation study of cold electron heating by loss cone instabilities, *Geophys. Res. Lett.*, 7, 889, 1980.

Ashour-Abdalla, M. and H. Okuda, Plasma physics on auroral field lines: The formation of ion conic distribution, *in* High-Latitude Space Plasma Physics, ed. by B. Hultqvist and T. Hagfors, Plenum Publ. Corp., New York, p. 165, 1983.

Aşhour-Abdalla, M. and H. Okuda, Electron acoustic instabilities in the geomagnetic tail, *Geophys. Res. Lett.*, 13, 366, 1986a.

Ashour-Abdalla, M. and H. Okuda, Theory and simulations of broadband electrostatic noise in the geomagnetic tail, *J. Geophys. Res.*, 91, 6833, 1986b.

Ashour-Abdalla, M., R.L. Richard, and F.V. Coroniti, A simulation study of electron hole distributions, *Geophys. Res. Lett.*, 16, 1137, 1989.

Auer, R.D., R.W. Kilb, and W.F. Crevier, Thermalization in the earth's bow shock, *J. Geophys. Res.*, 76, 2927, 1971.

Axford, W.I. and C.O. Hines, A unifying theory of high-latitude geophysical phenomenon and geomagnetic storms, *Can. J. Phys.*, 39, 1433, 1961.

Belmont,G., D. Fontaine, and P. Canu, Are equatorial electron cyclotron waves responsible for diffuse auroral precipitation?, *J. Geophys. Res.*, 88, 9163, 1983.

Birn, J., Three-dimensional computer modeling of dynamic reconnection in the magnetotail: Plasmoid signatures in the near and distant tail, *in* Magnetic Reconnection in Space and Laboratory Plasma, ed. by E.W. Hones, Jr., Geophys. Monogr. ser., AGU, Washington, DC, 30, p. 264, 1984.

Brice, N.M., Bulk motion of the magnetosphere, *J. Geophys. Res.*, 72, 5193, 1967.

Büchner, J. and L.M. Zelenyi, Deterministic chaos in dynamics of charge particles near a magnetic field reversal, *Phys. Lett. A.*, 118, 395, 1986.

Coroniti, F.V. and M. Ashour-Abdalla, Electron velocity space hole modes and narrowbanded electrostatic noise in the distant tail, *Geophys. Res. Lett.*, 16, 747, 1989.

Denavit, J. and J.M. Walsh, Nonrandom initialization of particle codes *in* Comments Plasma Phys. Controlled Fusion, vol. 6, pg. 209-223, 1981

Drummond, W.E. and M.N. Rosenbluth, Anomalous diffusion arising from microinstabilities in a plasma, *Phys. Fluids*, 5, 1507, 1962.

Dungey, J.W., Interplanetary magnetic field and the auroral zones, *Phys. Rev. Lett.*, 6, 47, 1961.

Dusenbery, P.B. and L.R. Lyons, The generation of electrostatic noise in the plasma sheet boundary layer, *J. Geophys. Res.*, 90, 10935, 1985.

Fredricks, R.W. and F.L. Scarf, Recent studies of magnetospheric electric field emissions above the electron gyrofrequency, *J. Geophys. Res.*, 78, 310, 1973.

Grabbe, C.L. and T.E. Eastman, Generation of broadband electrostatic noise by ion beam instabilities in the magnetic tail, *J. Geophys. Res.*, 89, 3865, 1984.

Gurnett, P.A., L.A. Frank, and R.P. Lepping, Plasma waves in the distant magnetotail, *J. Geophys. Res.*, 81, 6059, 1976.

Hones, E.W., Jr., Transient phenomena in the magnetic tail and their relationship to substorms, *Space Sci. Rev.*, 23, 393, 1979.

Hones, E.W., Jr., D.N. Baker, S.J. Bame, W.C. Feldman, J.T. Gosling, D.J. McComas, R.D. Zwickl, J.A. Slavin, E.J. Smith, and B.T. Tsurutani, Structure of the magnetotail at 220 RE and its response to geomagnetic activity, *Geophys. Res. Lett.*, 11, 5, 1984.

Kennel, C.F., F.L. Scarf, R.W. Fredricks, J.H. McGeeHee, and F.V. Coroniti, VLF electric field observations in the magnetosphere, *J. Geophys. Res.*, 75, 6136, 1970.

Kennel, C.F., F.L. Scarf, F.V. Coroniti, E.J. Smith, and D.A. Gurnett, Nonlocal plasma turbulence associated with interplanetary shocks, *J. Geophys. Res.*, 87, 17, 1982.

Kindel, J.M., Field aligned current instabilities in the high latitude boundary, Ph.D. thesis, University of California at Los Angeles, 1970.

Kindel, J.M. and C.F. Kennel, Topside current instabilities, *J. Geophys. Res.*, 76, 3055, 1971.

Kurth, W.S., L.A. Frank, M. Ashour-Abdalla, D.A. Gurnett, and B.G. Burek, Observations of a free energy source for intense electrostatic waves, *Geophys. Res. Lett.*, 7, 293, 1980.

Lembege, B. and R. Pellat, Stability of a thick two-dimensional quasi-neutral sheet, *Phys. Fluids*, 22, 1995, 1982.

Leroy, M.M., C.C. Goodrich, D. Winske, C.S. Wu, and K. Papadopoulos, Simulations of a perpendicular bow shock, *Geophys. Res. Lett.*, 8, 1269, 1981.

Levy, R.H., H.E. Petschek, and G.L. Siscoe, Aerodynamic aspects of the magnetospheric flow, *AIAA J.*, 2, 2065, 1964.

Mozer, F.S., C.A. Cattell, M.K. Hudson, R.L. Lysak, M. Temerin, and R.B. Torbert, Satellite measurements and theories of low altitude auroral particle acceleration, *Space Sci. Rev.*, 27, 155, 1980.

Ogino, T., R.J. Walker, and M. Ashour-Abdalla, Magnetic flux ropes in 3-dimensional MHD simulations, *Geophys. Monogr. Ser.*, submitted, 1989.

Pritchett, P.L., M. Ashour-Abdalla, and J.M. Dawson, Simulation of the current-driven electrostatic ion cyclotron instability, *Geophys. Res. Lett.*, 8, 611, 1981.

Pritchett, P.L., F.V. Coroniti, R. Pellat, and H. Karimabadi, Collisionless reconnection in a quasi-neutral sheet near marginal stability, *Geophys. Res. Lett.*, 16, 1269, 1989.

Rodriguez, P. and D.A. Gurnett, Electrostatic and electromagnetic turbulence associated with the Earth's bow shock, *J. Geophys. Res.*, 80, 19, 1975.

Roeder, J.L. and H.C. Koons, A survey of electron cyclotron waves in the magnetosphere and the diffuse auroral electron precipitation, *J. Geophys. Res.*, 94, 529, 1989.

Rönnmark, K., H. Borg, P.J. Christiansen, M.P. Gough, and D.J. Hones, Banded electron cyclotron harmonic instability - a first comparison of theory and experiment, *Space Sci. Rev.*, 22, 401, 1978.

Sagdeev, R.Z., Cooperative phenomena and shock waves, collisionless plasma, *in* Reviews of Plasma Physics, vol. 4, edited by M.A. Leontovich, pp. 23-90, Consultants Bureau, New York, 1966.

Scarf, F.L., F.V. Coroniti, C.F. Kennel, E.J. Smith, J.A. Slavin, B.T. Tsurutani, S.J. Bame, and W.C. Feldman, Plasma wave spectra near slow mode shocks in the distant magnetotail, *Geophys. Res. Lett.*, 11, 1050, 1984a.

Scarf, F.L., F.V. Coroniti, C.F. Kennel, R.W. Fredricks, D.A. Gurnett, and E.J. Smith, ISEE-3 wave measurements in the distant geomagnetic tail and boundary layer, *Geophys. Res. Lett.*, 11, 335, 1984b.

Schriver, D. and M. Ashour-Abdalla, Generation of broadband electrostatic noise: The role of cold electrons, *J. Geophys. Res.*, 92, 5807, 1987.

Walker, R.J., T. Ogino and M. Ashour-Abdalla, A global magneto-hydrodynamic model of magnetospheric substorms, *in* Physics of Space Plasma, edited by T. Chang, G.B. Crew and J.R. Jasperse, Scientific Publishers, Cambridge, MA, SPI Conference Proceedings and Reprint Series, 7, p. 235, 1988.

Walker, R.J. and T. Ogino, Global magnetohydrodynamic simulations of the magnetosphere, *IEEE Transactions on Plasma Science,* 17, 2, 135, 1989.

List of Participants in

the Crafoord Symposium

Stockholm, September 28-29, 1989

S-I. Akasofu	University of Alaska	Fairbanks
H.Alfvén	Royal Institute of Technology	Stockholm
M. Ashour-Abdalla	University of California	Los Angeles
A. Bahnsen	Danish Space Research Institute	Lyngby
D.N. Baker	Goddard Space Flight Center	Greenbelt
P. Bauer	Service d'Aeronomie du CNRS	Verrieres-le-Buisson Cedex
L. Block.	Royal Institute of Technology	Stockholm
R.M Bonnet	European Space Agency	Paris
R.Boström	Swedish Institute of Space Physics	Uppsala
A. Brekke	University of Tromsö	Tromsö
D.A Bryant	Rutherford and Appleton Lab.	Chilton, Didcot
J.L. Burch .	Southwest Research Inst.	San Antonio
C.R Chappell	Marshall Space Flight Center	Huntsville
P. Christiansen	University of Sussex	Brighton
A. Egeland	University of Oslo	Blindern, Oslo
D.S. Evans	NASA Hq	Washington D.C.
L.A. Frank	University of Iowa	Iowa City
K. Fredga	Swedish National Space Board	Solna
T.A Fritz	Los Alamos National Lab	Los Alamos
C.-G.Fälthammar	Royal Institute of Technology	Stockholm
K. Gringauz	Space Research Institute	Moscow
D.A. Gurnett	University of Iowa	Iowa City
G. Gustafsson	Swedish Institute of Space Physics	Uppsala
G. Haerendel	Max-Planck-Institut für Extraterr. Phys.	Garching
W.J. Heikkila	The University of Texas at Dallas	Richardson
B. Hultqvist	Swedish Institute of Space Physics	Kiruna
A.D. Johnstone	University of London	Dorking
C. Kennel	University of California	Los Angeles
S.M. Krimigis	Applied Physics Lab.	Laurel
L.J. Lanzerotti	AT & T Bell Laboratories	Murray Hill
L. Liszka	Swedish Institute of Space Physics	Umeå
J.G. Luhmann	UCLA Space Science Center	Los Angeles
R. Lundin	Swedish Institute of Space Physics	Kiruna
R. Lüst	ESA Hq	Paris
B.N Maehlum	Försvarets Forskningsinstitutt	Kjeller
G. Marklund	Royal Institute of Technology	Stockholm
C.E. McIlwain	Univ. of California at San Diego	La Jolla
V.V. Migulin	IZMIRAN	Moscow
V.M. Mishin	Sibizmir	Irkutsk
F. Mozer	University of California	Berkely
J.S. Murphree	University of Calgary	Calgary
N.F. Ness	University of Delaware	Newark
J. Nishimura	ISAS	Tokyo
A. Pedersen	ESTEC	Noordwijk
R. Pellinen	Finnish Meteorological Institute	Helsinki
O. Pokhotelov	Institute of Physics of the Earth	Moscow
E.A. Ponomarev	Sibizmir	Irkutsk

T.A. Potemra	Applied Physics Lab.	Laurel
H. Reme	University of Toulouse	Toulouse
J.G. Roederer	University of Alaska	Fairbanks
G. Rostoker	University of Alberta	Edmonton,
A. Roux	CNET/CRPE	Issy-les-Moulineaux
C.T. Russell	UCLA Space Science Center	Los Angeles
K. Rönnmark	Swedish Institute of Space Physics	Umeå
R. Sagdeev	Space Research Institute	Moscow
S.D. Shawhan	NASA Hq	Washington
E.G. Shelley	Lockheed Missiles and Space Comp.	Palo Alto
G.G. Shepherd	York University	Downsview
L. Stenflo	University of Umeå	Umeå
J. Stiernstedt	Swedish National Space Board	Solna
H. Strub	Der Bundesministerium für Forschung und Technologie	Bonn
F. Söraas	University of Bergen	Bergen
P. Tanskanen	University of Oulu	Oulu
E. Ungstrup	Danish Space Research Institute	Lyngby
J.A. Van Allen	University of Iowa	Iowa City
M. Walt	Lockheed Missiles & Space Comp.	Palo Alto
B.A. Whalen	Herzberg Institute of Astrophysics	Ottawa
D.J. Williams	Applied Physics Lab.	Laurel
J.D. Winningham	Southwest Research Institute	San Antonio

INDEX